# The Organization of Heredity

Kenneth R. Lewis
and Bernard John

CONTEMPORARY BIOLOGY

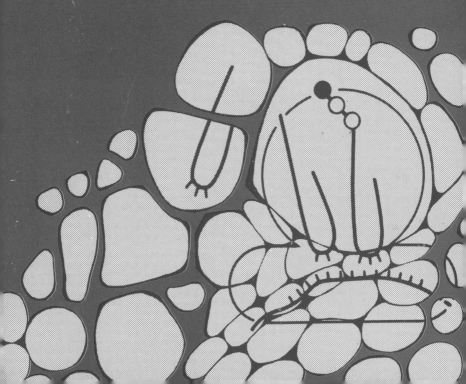

# CONTEMPORARY BIOLOGY

Spectacular progress in biological research in recent years has led to many changes in the form and content of the biological curriculum. Advances in molecular biology and comparative studies in many fields have emphasized the importance of the unity of life as well as its diversity. This series of student texts has been designed against this background.

ALREADY PUBLISHED

The Biology of Fungi, Bacteria and Viruses 2nd edition.   *Greta Stevenson*
The Biology of Lichens   *Mason E. Hale, Jr.*
Principles of Animal Physiology   *Dennis W. Wood*
Statistics and Experimental Design   *Geoffrey M. Clarke*
The Comparative Endocrinology of the Invertebrates

*Kenneth C. Highnam and Leonard Hill*
The Physiology of Flowering Plants   *H. E. Street & H. Öpik*
The Organization of Heredity   *Kenneth R. Lewis & Bernard John*
The Life and Organization of Birds   *W. B. Yapp*

IN PREPARATION

The Biology of the Arthropoda   *Kenneth U. Clarke*
Molecular Architecture and Cellular Function   *David J. D. Nicholas*
Ecology   *Amyan Macfadyen & Gordon Goodman*
The Micrometeorology of Plants and Animals   *John L. Monteith*
Theories of Development   *Max Hamburgh*
The Biology of the Protozoa   *Michael A. Sleigh*
The Biology of Parasitism   *John A. Clegg*

*A series of student texts in*

# CONTEMPORARY BIOLOGY

---

*General Editors:*

Professor E. J. W. Barrington, F.R.S.
Professor Arthur J. Willis

' If these d'Herelle bodies were really genes, fundament-
ally like our chromosome genes, they would give us an
utterly new angle from which to attack the gene problem.
They are filterable, to some extent isolable, can be handled
in test tubes, and their properties,´ as shown by their
effects on bacteria, can be studied after treatment. It
would be very rash to call these bodies genes, and yet at
present we must confess that there is no distinction known
between genes and them. Hence we cannot categorically
deny that perhaps we may be able to grind genes in a
mortar and cook them in a beaker after all. Must we
geneticists become bacteriologists, physiological chemists
and physicists, simultaneously with being zoologists and
botanists ? Let us hope so.'

H. J. MULLER (1922)

(from Variation due to change in the individual gene,
*Am. Nat.* **56**, 32–50)

# The Organization
# of Heredity

Kenneth R. Lewis
M.A., Ph.D.
University Lecturer and Fellow of Exeter College, Oxford

and

Bernard John
Ph.D., D.Sc.
Professor of Zoology, University of Southampton

American Elsevier Publishing Company, Inc.
New York

122314

*576.5*
*L 674*

First published 1970

American Elsevier Publishing Company, Inc.
52 Vanderbilt Avenue, New York, N.Y. 10017

First published in Great Britain by
Edward Arnold (Publishers) Ltd.

Standard Book Number (Cloth edition): 444-19637-4
Standard Book Number (Paper edition): 444-19638-2
Library of Congress Catalog Card Number: 70-130964

Printed in Great Britain by
William Clowes and Sons, Limited, London and Beccles

# Preface

Heredity is the essence of life and an understanding of it is crucial in the study of a wide range of biological and parabiological problems.

Little advance was made towards elucidating the phenomena of heredity and variation until Mendel's conclusions were confirmed, extended and publicly debated following their rediscovery. However, the progress made during the first half of this century was very rapid and the advances of the last twenty years have transformed the subject completely. These years have seen the identification of the genetic material, its *in vitro* synthesis and a determination of its molecular structure. Further, the synthetic pathways by which the genotype gains phenotypic expression have been largely defined and the code which connects the language of nucleic acid with that of protein has been revealed. Practically all these advances came from studies on viruses, bacteria and fungi which are rapidly replacing maize, mice and *Drosophila melanogaster* as the classical materials of experimental genetics.

This book deals, almost exclusively, with recent advances in molecular and microbial genetics at a level appropriate to an introductory course in these important and advancing areas of enquiry. The methods used for mapping mutant sites in various organisms are considered in some detail because they are relevant to the investigation of numerous aspects of genetic organization. These include the mechanism of crossing-over, the nature of the genetic code and the functional co-ordination of the genotype which are also discussed.

We hope that this account of a molecular and analytical approach to the study of heredity will enable the reader to realize the essential unity of life and to apply the principles revealed to the study of higher levels of biological organization.

K. R. Lewis,
Botany School,
University of Oxford
1970

B. John,
Zoology Department,
University of Southampton

# Table of Contents

# *Prelude*

## A BIOLOGICAL BACKGROUND

A living organism is a highly ordered system existing in a comparatively unordered environment. The system does not merely maintain itself but actually expands and extends at the expense of the environment (Development and Differentiation). It does so by taking in matter from the environment (Ingestion and Absorption) and imposing its own kind of order on it. The materials which the environment can offer are not always in a form in which they can be absorbed by the organism. In this event the organism itself must first change these materials into other, even less ordered entities prior to their uptake (Digestion).

This process of degradation and, even more so, the subsequent re-ordering require energy which, again, must be derived ultimately from the environment either in chemical form or as light. Whatever the source of this capacity for doing work, it can be stored in the form of chemical energy. The controlled break-down of these energy-containing compounds (Katabolism) can then provide the energy required for maintenance, repair and development (Anabolism).

Of course, the organism cannot create matter or energy, and the conversions of which it is capable, while numerous, are limited. What is more, they vary between organisms so that an environment which is tolerable or even favourable for one may be totally unsuitable for another. Thus, the diversity of organisms is related to the diversity of environments (Ecology).

Even the environment of one individual is heterogeneous, even at one time. At a gross level, there is air and earth, and an organism may respond to this heterogeneity by producing upgrowing shoots and descending roots. Thus, an organism does not simply exploit its environment; it is designed to exploit it (Adaptation). Further, it not only translates the environment; it interprets the environment. This must mean an awareness of its circumstances and response to them (Irritability). We must be very careful, however, to distinguish those features which prove to be adaptive to a particular situation but develop even outside that situation, from those features, adaptive or otherwise, which are generated only in relation to a particular circumstance.

We can see, therefore, that because the environment sustains the organism, it must also be changed by the organism. And in this way one living system can alter, and be held to form part of, the environment of other living systems. This interaction, or communication via environmental modification, may be competitive or co-operative.

The physiological, developmental and ecological aspects outlined above relate to the prospects and problems of life for the individual. But individuals die. There comes a time when, for no apparent reason, the complete organism can no longer resist the physical forces which tend to create disorder. But while it, as an individual, must give up the fight it does not 'go gentle into that good night'. It can produce and isolate parts of itself which are able, in their turn, to expand and extend at the expense of the environment. Is this neophyte as capable as its progenitors in exploiting the same environment, or a different one? This will depend, in part, on whether the parent produced the propagulum unaided or whether it co-operated in this venture with another of its own kind (Mating System).

Thus, the properties of living systems are diverse and diversified but on what do they depend? So far we have talked only of properties and not of construction but, clearly, function depends on form and can sometimes be inferred from it. To understand how living systems work therefore, we must study their design.

They are in fact composed of the common elements of the inorganic world as, of course, they must be because this is their ultimate source of supply. Within the organism these exist in a simple form, to some extent, but for the most part they are compounded into molecules which rarely, if ever, occur outside living systems or their derivatives (milk, manure). This much is true, however, not only of a living organism but also of a dead one, for the organism, like a machine or a legislature, must be regarded as a hierarchy of organizations. Thus, it is not the molecules as such which give living systems their unique properties but the higher levels of integration into which they are organized.

An organism is, in fact, fashioned from multi-molecular membranes, threads and particles which are themselves arranged in a more or less consistent pattern within one or more cells. And the properties discussed earlier arise from the complete organization for here, as elsewhere, the whole is greater than the sum of its parts.

There is a sense in which this consistent construction is misleading, as it can be at higher levels of organization, for while the construction itself appears more or less constant and, indeed, is constant in that its over-all molecular composition varies only between very narrow limits, most of the actual individual molecules, and, indeed, some cells, are subject to periodic break-down and replacement by others of their own kind.

Thus, the organism owes its life to its over-all organization and the above properties are not all shown, or not all shown equally, by each of its parts. This mosaicism of structure and function creates a mutual dependence which leads to, and yet depends on, mutual adjustment and adaptation. This situation occurs also at higher levels of biological organization and not only within individuals but beyond them. Thus, males and females are mutually dependent and co-adapted; they need each other as much as nuclei need cytoplasm and vice versa. Likewise, queens and drones are fed by the workers they beget.

Thus, there is no clear lower limit to biological organization and no well-defined upper limit either. But irrespective of the level of organization, this constant-by-change, stable-through-substitution system increases, not by the mere extension of existing parts but by a multiplication of these parts. Thus, development is a kind of repro-duction.

How, then, are the new parts produced? Does A give rise to A and B to B, or does A produce both A and B? Does A produce B while B produces A, or do they both come from C? There are various *a priori* possibilities but the general question remains—does the production of some parts depend on the prior existence of similar, or dissimilar, parts, or can they be created *de novo*? Clearly, only a tree can make a tree and only a cell can make a cell. But how far does this hereditary hierarchy extend? Let us see.

# PART I

THE CHEMICAL ORGANIZATION
OF THE GENOTYPE

# I

# *The Material Basis of Heredity*

## THE ORGANELLAR BASIS OF HEREDITY

Mendel's breeding experiments with the garden pea provided the basis for 20th century research in Genetics. But although these experiments were performed, reported and published by 1866 they were not widely known until after the turn of the century. Thus, the research of the 19th century was undertaken in ignorance of Mendel's critical approach and fundamental discoveries. It can be fairly claimed, however, that the cytological studies of the nineteenth century provided strong indications that the cell nucleus was the principal carrier of heredity.

The term nucleus was applied by Brown in 1833 to the more or less spherical body he had observed in a variety of plant cells including pollen grains. Over 40 years later, again from a study of plant cells, Strasburger concluded that nuclei did not arise *de novo* but were produced only by the division of pre-existing nuclei. At about the same time, studies by Hertwig and others on fertilization in animals showed that gametic fusion involved also a fusion of nuclei.

In 1883, van Beneden working on the thread worm, *Parascaris equorum*, showed further that while the nuclei of both egg and sperm each contained two threads, the zygote nucleus contained four. These threads, which Waldeyer called chromosomes in 1888, had been seen earlier but they proved to be visible only in dividing cells. In 1882 Flemming gave the name mitosis to the process of somatic cell division. He himself had carefully studied this sequence in salamander larvae. He found that the mitotic cycle consisted essentially of the accurate longitudinal division of the chromosomes followed by a separation of the half-chromosomes or chromatids. In this way two daughter nuclei were produced which had the same complement of chromosomes as each other and the original cell. Further, in 1888 Boveri showed that although the chromosomes could not be individually resolved in non-dividing nuclei, they appeared at the beginning of one division in the same positions they occupied at the end of the previous division. This indicated that they maintained their integrity even when they could not be seen.

If then the mitotic cycle effected a regular perpetuation of the chromosome complement, how was it that the gametes of *Parascaris* had only half as many chromosomes as the zygotes?

It had been found previously that the formation of gametes in animals and of the equivalent spores in plants was generally preceded by two divisions of the nucleus which followed each other in rapid succession. The first of these did not resemble the standard mitotic pattern, and the halved number of chromosomes appeared at the second division. It was also known that the division sequence in the egg mother cells (oocytes) of sexually reproducing animals was different from that in the oocytes of parthenogenetic forms. In the former, the first division was followed by the extrusion of one daughter nucleus with the formation of a so-called polar body. The remaining nucleus divided a second time and one of the products of this division also was extruded, as a second polar body. Thus, each mother cell gave rise to one functional egg. In parthenogenetic forms, on the other hand, where the nuclear fusion of fertilization is omitted, only one polar body was formed. Thus, the retention of one polar body appeared to make-up, as it were, for the absence of the sperm nucleus.

On evidence such as this, Weismann in 1887 predicted that at some stage (e.g. gamete formation) in each sexual cycle, a special kind of nuclear division occurred which led to a halving of the chromosome number. It is now known that the double-division indicated above performs this and other functions. It was studied in some detail by Winiwarter in the ovaries of rabbits in 1900 and was given its present name of meiosis by Farmer and Moore five years later.

These cytological studies of the last century indicated that the nucleus or, more particularly, the chromosomes, themselves possessed the property of heredity which suggested that they were also responsible for heredity. Clearly, the evidence was indirect. Heredity is generally recognized and frequently studied, then as now, by comparing the characteristics of offspring with those of their parents. Clearly, therefore, the role of the nucleus in heredity would have been more acceptable if it had been shown to have a role in development. This demonstration was provided by Boveri in 1889.

Prior to this demonstration, however, Weismann had already proposed his theory of what is generally called the continuity of the germ plasm. It had many important features not least of which was the utter rejection of the adaptive environmental modification of heredity. But the aspect of the theory which concerns us here is the basic notion which, amongst other things, led to this rejection, namely, the view that

'Heredity is brought about by the transference from one generation to another, of a substance with a definite chemical, and above all, molecular constitution.'*

What then is this chemical and what is its molecular constitution? It was first isolated by Miescher in 1868 from the nuclei of pus cells and salmon sperm and it was given its present name by Altmann in 1889. But over fifty years elapsed before its role was appreciated and its detailed structure elucidated.

## THE MOLECULAR BASIS OF HEREDITY

In 1928 Griffith encountered a phenomenon now known as genetic transformation.[41] Colonies of virulent strains of the pneumococcus *Diplococcus pneumoniae*, grown on nutrient agar, have a smooth (S), glistening appearance owing to the presence of a type-specific, polysaccharide capsule which surrounds the cells. Avirulent strains, on the other hand, lack this capsule and they produce dull, rough (R) colonies. Smooth forms do mutate to the rough at low frequency but this change does not appear to be reversible. Further, a change from one virulent type to another does not appear to occur during the comparatively short periods of experimental observation.

Griffith found that neither heat-killed virulent cells nor living avirulent cells, when they were injected separately, had any effect. But when they were injected together the recipient mice frequently became infected. Further, living, encapsulated, virulent cells could be isolated from such mice.

Clearly, since the resurrection of the heat-killed S cells could be safely discounted, the only conclusion was that the living R cells in the original mixture, or their products, had been somehow transformed into virulent forms following an association with dead S cells. Griffith used virulent

* Weismann, A. (1885). The continuity of the germ-plasm as the foundation of a theory of heredity. Authorized English translation in *Essays upon Heredity and Kindred Biological Problems*, ed. E. B. Poulton, S. Schonland and A. E. Shipley, Clarendon Press, Oxford.

strains of serotype III and avirulent strains derived from serotype II and found that transformation was always to the virulent type of the dead S cells. Further, the transformations were stable and the effect was transmitted to the daughter cells over innumerable generations. Four years later, Alloway showed decisively that transformation could be effected *in vitro* by adding filtered extracts of heat-killed S cells to liquid cultures of R strains. He did not identify the 'activating stimulus of a specific nature' but he did show that the purified capsular material itself was not effective.

Then, in 1944, Avery, MacLeod and McCarty published the very important discovery that a cell-free and highly purified deoxyribonucleic acid (DNA) extract could bring about transformation.[6] The yield of transformed cells was low under these circumstances, however, only about 1 in $10^6$ cells being changed.

At this time protein was generally accepted as being the genetic material and so considerable care was taken to identify the active agent. Thus, serological tests did not reveal any polysaccharide, and proteolytic enzymes had no effect on the transforming activity of the extract. But this activity was completely destroyed by deoxyribonuclease (DNase). This did not convince everybody, but when Hotchkiss showed that DNA contaminated with less than 0·02% protein was effective, the nature of the transforming principle was no longer in doubt.[54]

The transformation of several traits has now been achieved in a number of bacteria (Table 1.1). Yields of about 1 in $10^2$ can be obtained and the method is a very useful technique in genetic research on bacteria (see p. 91). But, although there are claims to the contrary, there is no real evidence of transformation in higher organisms. Two structural features of the bacterial cell are of importance in this connection. First, the DNA in bacteria and blue-green algae, unlike that in the chromosomes of higher organisms, is not associated with protein. Second, the 'nuclear' DNA is not confined within a nuclear membrane as it is elsewhere.

The means by which the transforming DNA enters the cell is not known with certainty but from the above statements it is clear that, once it is inside the bacterial cell, the recipient genome is more accessible to its influence than it would be in other organisms.[67]

The hereditary significance of DNA was further demonstrated in 1952 by Hershey and Chase who studied viruses which infect bacteria. These viruses were discovered by D'Herelle in 1917 who found that filter-passing agents, invisible by ordinary microscopy, entered and multiplied within the bacterial host and were subsequently released by the rupture (lysis) of the bacterial cell.

A variety of these so-called bacteriophages has now been studied with regard to structure, chemical composition and genetic organization.

**Table 1.1**  Hereditary characters which are transformable in bacteria. (After Braun, W. (1965) *Bacterial Genetics*, 2nd edn., W. B. Saunders Co., Philadelphia and London.)

| Character | Species |
| --- | --- |
| Capsular polysaccharide synthesis | *Diplococcus pneumoniae* |
| | *Hemophilus influenzae* |
| | *Neisseria meningitidis* |
| | *Escherichia coli* |
| | *Xanthomonas phaseoli* |
| Filamentous type of growth | *D. pneumoniae* |
| Specific protein antigens | |
|   M protein | *D. pneumoniae* |
| Drug and antibiotic resistance | |
|   Penicillin: 3 levels | *D. pneumoniae* |
|   Streptomycin: several levels | *D. pneumoniae* |
|   Streptomycin: high level | *H. influenzae, N. meningitidis, X. phaseoli* |
|   Sulphanilamide: several levels | *D. pneumoniae* |
|   Erythromycin: several levels | *D. pneumoniae* |
|   Erythromycin | *N. meningitidis* |
|   Bryamycin, Canavanine, Aminopterin, Amethopterin, 8-Azaguanine | *D. pneumoniae* |
| Antibiotic dependence | *N. meningitidis* |
| Synthesis of specific enzymes | |
|   Mannitol dehydrogenase, Maltase, Lactic acid oxidase | *D. pneumoniae* |
|   Sucrase, β-galactosidase | *B. subtilis* |
| Sporulation | *B. subtilis* |
| Ability to infect plants | *Agrobacterium* sp. |
| | *Rhizobium* sp. |

Certain phages which attack various forms of the colon bacillus, *Escherichia coli* strain B, are labelled T phages and they are numbered according to host range and other properties. The phage used by Hershey and Chase was $T_2$, the structure of which is more or less typical of the T group.[12] This virus is composed of about 40% by weight of DNA and 60% protein with small quantities of lipid. The DNA is contained in the head, and the proteins of the head membrane; the sheath and the tail fibres differ from one another (Fig. 1.1).

The enzyme DNase degrades DNA but the DNA of intact viruses is not susceptible to attack owing to the protection afforded by the head membrane. However, the DNA and head protein are not firmly conjugated in the form of nucleoprotein. Thus, if a strong aqueous solution of salt containing a suspension of viruses is rapidly diluted, the particles

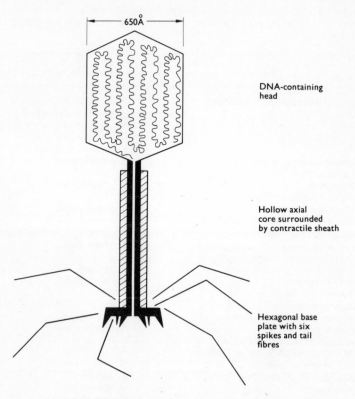

Fig. 1.1   The structure of $T_2$ phage drawn approximately to scale. The diameters of the axial hole and the tail fibres are 25 Å and 20 Å respectively. An axial sheath has not been found in $T_5$ and lambda which have thinner, more flexible tails with a single spike and no tail fibres.

suffer what is called an 'osmotic shock'. As a result of this the DNA is released into the medium and, having lost its protective envelope, becomes vulnerable to DNase degradation. The DNA-protein separation effected by the shock is reasonably clean. Thus, when the protein fraction, which does not contain phosphorus, is labelled with $^{35}$S and the DNA, which does not contain sulphur, is tagged with $^{32}$P then, following the shock, radio-sulphur is virtually confined to the 'ghosts' and the radio-phosphorus to the DNA fraction. The protein ghost retains its original structure more or less intact and it retains also many of the properties of the original, complete virus. For example, the infection of bacteria by intact virus is initiated by the attachment of the viral tail to the bacterial cell wall. The tail-end is occupied by a hexagonal plate with six short spikes and as many tail fibres. This region appears to determine the host range of the virus. It would further appear that the distal tail region possesses also the lysozyme activity which enables the bacterial wall to be penetrated. The lysozyme of these viruses is capable of attacking mucopolysaccharides, mucoproteins and chitin. In fact it has properties very similar to those of the lysozyme present in egg-white.

Support for the localization of these functions in the protein of the tail-end region is seen in the fact that DNA-free ghosts and even detached tail fibres are adsorbed by bacteria and they show the same host specificity as the intact virus from which they came. Attachment by ghosts may even lead to the death and lysis of the bacterium.

What then is the function of the DNA? This question was answered by Hershey and Chase.[51] As in the osmotic shock experiment, they labelled the viral protein with $^{35}S$ and the DNA with $^{32}P$. Labelled viruses were then allowed to infect susceptible hosts. Unattached viruses were removed by centrifugation and the suspension of bacterial cells with adsorbed viruses was vigorously agitated in a blendor. This treatment detaches those parts of the virus which normally remain attached to, but outside, the bacterial cell following infection. At least 80% and as much as 97% of the $^{35}S$ was found to be in this fraction. Most of the $^{32}P$-labelled DNA, on the other hand, was found to be inside the bacterial cell following infection. It appeared, therefore, that the process of infection was rather like the osmotic shock in that the DNA was injected into the bacterial cell while the major protein fraction remained outside.

Though agitated, the bacterial cells lysed and released a crop of normal phage particles with the properties of the parental strain. Over 40% of the radio-phosphorus in the parental phage could be traced to the offspring but little or no $^{35}S$ could be detected in them. Ghost 'infected' bacteria, on the other hand, while they may lyse, yield no phage.

It is worth noting that the degree of contamination in this experiment was considerably higher than that in the transformation experiments. It is very likely, therefore, that the reception afforded the experiments of Hershey and Chase would not have been as favourable had it not been for the prior demonstration of DNA transformation.

It is worth noting also that even the transformation experiments roused little interest when they were first published. The widespread indifference stemmed largely from the fact that the bacteriological literature is full of examples of phenotypic modification. Some bacteria produce glistening colonies of cells with well-developed capsules on media containing sucrose, but when glucose is used as a carbon source matt colonies of cells with thin capsules are produced. The distinctive feature of transformation by DNA, therefore, was its stable and genetic nature. But in organisms with genetic systems as strange as those of bacteria it is not always easy to demonstrate unambiguously the genetic nature of a change.[24] In fact, until 20 years ago, bacteria had no genetics! Indeed this is still true of certain groups such as the blue-green algae.

These experiments, like all important ones, posed new problems.

Prior to them it had been generally assumed that genetic functions were performed by proteins, the known variety and specificity of which seemed to fit them for this role. But the above experiments left little doubt that DNA without conspecific protein could contain and transmit all the information required to specify the nature of new virus particles. In other words, DNA showed heredity and determined development.

It will be appreciated, however, that a demonstration of the genetic role of DNA does not necessarily mean that all DNA performs a genetic function or that DNA is the sole genetic material. In fact, the knowledge that most plant viruses, like some of those which attack animals or bacteria, do not contain DNA was sufficient evidence that genetic functions were not a prerogative of this nucleic acid.[37] The genetic role of a similar acid which contains ribose sugar was convincingly demonstrated by Fraenkel-Conrat in 1956.[33]

He worked on the tobacco mosaic virus which is composed of ribonucleic acid (RNA) surrounded by a cylinder of protein about 300 m$\mu$ long and 15 m$\mu$ wide (Fig. 1.2). Here too the protein can be separated from the RNA. Gierer and Schramm showed that the isolated RNA could infect the host and give rise to normal progeny (cf. transformation, p. 8). But in this respect it was only about 0·1% as efficient as the intact virus, probably owing to its digestion by RNase during the process of infection. Further, it was found that the small, identical molecules which form the protein component could re-aggregate in the absence of RNA to produce short rods very similar in appearance to the intact virus. These reconstituted 'ghosts', however, could not produce disease symptoms. Re-aggregation is much more efficient in the presence of RNA and it leads to the reconstruction of apparently normal, infective virus particles.

Fraenkel-Conrat produced chimaeral viruses by allowing the protein blocks derived from one strain to re-aggregate in the presence of RNA obtained from a different strain. The two strains employed (Nicotiana and Plantago) differ appreciably and they are clearly distinguishable on the basis of disease symptoms, amino acid composition and serological properties.

The chimaeras were found to have the serological character of the protein donor (cf. bacteriophage ghosts). But the disease symptoms produced by the chimaera following its reproduction were those of the

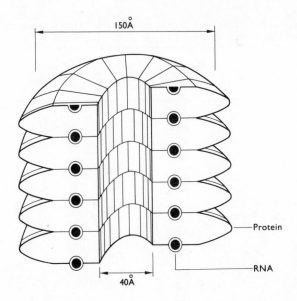

**Fig. 1.2** A longitudinal section of a short length of the rod-shaped tobacco mosaic virus. The single-stranded RNA is held in a flat spiral (23 Å pitch) by the protein sub-units which are arranged helically around a hollow axis 40 Å in diameter.

RNA donor. The symptoms actually result from the production of progeny virus in the host cells. What is more the progeny showed the serological character of the RNA-donating strain and the amino acid composition of their protein followed this strain as well (Table 1.2).

Comparable chimaeras can arise spontaneously when a bacterium is simultaneously infected by two related phage strains. Each strain replicates its DNA and directs the synthesis of its specific protein. But when the protein-DNA complex of the progeny phage is produced, the components may 'hybridize'. Indeed, bacteriophages contain more than one kind of protein, and after mixed infection a given particle may be composed of a mixture of the parental-type protein. Notice that this so-called phenotypic mixing does not require recombination between the genomes.

Like so many discoveries, the demonstration of the genetic role of RNA in plant viruses raised its own problems. Viruses seemed to consist almost exclusively of protein and one or other of two kinds of nucleic acid, DNA and RNA. Genetic functions were a peculiar property of the latter. But bacteria and cellular plants and animals invariably contain

**Table 1.2**   The amino acid composition of two strains of TMV and of reciprocal chimaeras of their RNA and protein components (After Fraenkel-Conrat, H. L. and Singer, B. (1957) *Biochim. biophys. Acta*, **24**, 540–8.)

| Virus strain | RNA | Nicotiana | Nicotiana | Plantago | Plantago |
|---|---|---|---|---|---|
| | Protein | Nicotiana | Plantago | Nicotiana | Plantago |
| AA composition (% weight) of progeny protein | gly | 2·3 | 2·3 | 1·8 | 1·6 |
| | ala | 6·5 | 6·9 | 8·5 | 8·5 |
| | ser | 9·0 | 8·8 | 8·1 | 8·1 |
| | thre | 8·9 | 8·9 | 7·5 | 7·2 |
| | cyst | . . . not determined . . . . | | | |
| | meth | 0·0 | 0·0 | 2·2 | 2·0 |
| | val | 9·6 | 9·0 | 6·3 | 5·9 |
| | leu isoleu | 14·2 | 14·3 | 12·2 | 12·2 |
| | asp A asparag | 13·8 | 14·2 | 14·8 | 15·0 |
| | glut glut A | 12·4 | 12·1 | 17·3 | 16·4 |
| | arg | 9·5 | 9·7 | 8·5 | 8·9 |
| | lys | 1·9 | 1·8 | 2·3 | 2·4 |
| | phenyl | 7·2 | 7·1 | 5·4 | 5·3 |
| | tyr | 4·1 | 4·3 | 6·2 | 6·3 |
| | his | 0·0 | 0·0 | 0·7 | 0·7 |
| | trypt | 2·8 | 2·6 | 2·2 | 2·2 |
| | pro | 5·0 | 5·1 | 5·1 | 5·0 |

both DNA and RNA and although by far the greater fraction of DNA is confined to the nucleus, RNA is present in both nucleus and cytoplasm, predominantly (c. 90%) in the latter. Further, while the majesty of nuclear heredity cannot be denied, evidence for some extrachromosomal hereditary transmission has been accumulating since the early indications of it in the studies of Correns on the chloroplasts of plants (see p. 78). Furthermore, RNA is known to participate in protein synthesis which, in the case of the green alga *Acetabularia*, can continue in the cytoplasm for a considerable time after the removal of the nucleus.[44] What then is the position in higher organisms with regard to DNA, RNA and genetic function?

Higher organisms are not as easy to manipulate as bacteria and viruses but there can be little doubt that the principal if not the sole genetic material in them is DNA. For example, in 1950 Swift found that the amount of DNA per resting nucleus in mitotically-active tissues of animals ranged from twice (2C) the value found in the gametes (1C) to four times (4C) this value. He concluded that DNA synthesis occurred during the resting stage (interphase) between mitotic divisions. In this respect the behaviour of DNA paralleled that of the chromosomes. Mirsky too, from estimations of DNA content in a range of tissues from a variety of organisms, found a close similarity between the behaviour of chromosomes and DNA which was not paralleled by the RNA fraction. Likewise, in 1951 Howard and Pelc came to the conclusion that DNA synthesis in the root tips of the broad bean, *Vicia faba*, was confined to a part of interphase, and in the following year this situation was confirmed for tissue culture cells of amphibia, birds and mammals by Walker and Yates.

Different methods of studying the behaviour of the DNA were used in these investigations. Swift determined the relative DNA values by measuring photometrically the intensity of the colouration obtained following staining by the Feulgen method. This staining technique is specific for DNA and rests on the Schiff colour reaction given by the aldehydes which are liberated from DNA following mild acid hydrolysis. Walker and Yates, on the other hand, measured the absorption by the nuclei of ultra-violet light of wavelength 265 m$\mu$, while Howard and Pelc used radioactive $^{32}$P in conjunction with autoradiography (see p. 27).

In marked contrast to cellular RNA, which turns over rapidly, the genetic DNA is metabolically very stable and the incorporation of radio-active precursors into it is virtually confined to the interphase period of net DNA synthesis (so-called S phase). This too supports its genetic role as does evidence presented later.

# 2

# Nucleic Acid Structure and Synthesis

## DEOXYRIBONUCLEIC ACID

### The structure of DNA

A standard technique in the analysis of large molecules is to study the components released when the macromolecule is progressively broken down. The hydrolysis products of DNA were studied by Kossel and others as long ago as the last century. When DNA is hydrolysed by acid or enzyme treatment, it is degraded in ways which suggest the following construction:

DNA
↓
Nucleotides
↓
Nucleosides + Phosphoric acid
↓
Heterocyclic bases + Deoxyribose

As the sequence indicates, the glycosidic combination of the sugar with a base is called a nucleoside, and this in combination with phosphate is called a nucleotide. Thus, DNA consists of polynucleotide chains which electron microscopy has shown to be unbranched.

It was 1929 before it was finally agreed that all the sugar units in DNA are of one kind, namely, 2-deoxy-D-ribose. The name indicates the absence of the oxygen atom which is attached to the carbon atom numbered 2 ($C^2$) in D-ribose, the sugar present in RNA. In fact, the sugar exists as a ring in which four carbon and one oxygen atoms are involved. The carbons in the ring are numbered 1 to 4 in sequence from the oxygen so that the carbon outside the ring is number 5. The absence of an oxygen attached to $C^2$ means that deoxyribose contains only three hydroxyl groups and these occur at $C^1$, $C^3$ and $C^5$ (Fig. 2.1).

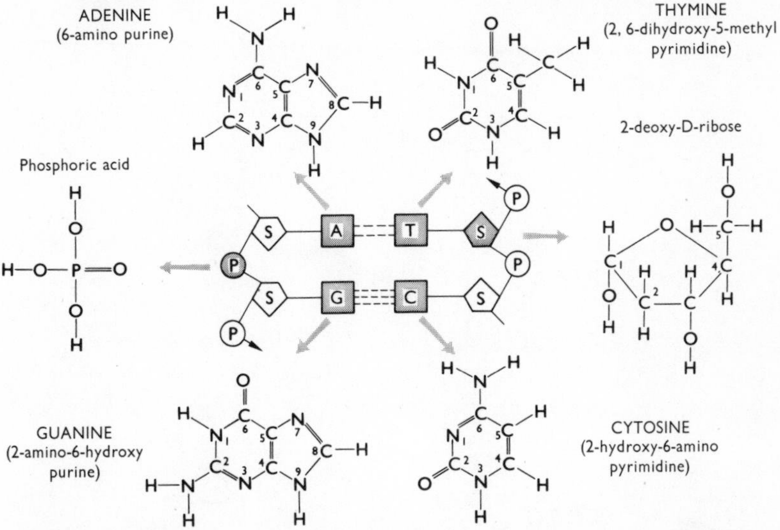

**Fig. 2.1** The molecular components of DNA and their general arrangement in the duplex structure. Two systems of numbering the pyrimidine ring are in current use. The older system, widely employed in Europe, is adopted here. The two systems are related as follows:

| Old | 1 | 2 | 3 | 4 | 5 | 6 |
|-----|---|---|---|---|---|---|
| New | 3 | 2 | 1 | 6 | 5 | 4 |

The nitrogenous bases are of two main types, double-ring purines and smaller, single-ring pyrimidines. In most samples of DNA only two kinds of purines, adenine and guanine, and two kinds of pyrimidines, thymine and cytosine, are found, or, at least, these are the bases which predominate (see p. 40).

In a nucleoside, the hydroxyl group $C^1$ in the sugar is involved in a glycosidic linkage with the $N^3$ of a pyrimidine or else with the $N^9$ of a purine. If then these mononucleosides are linked by the formation of phosphate diester bridges involving the two remaining hydroxyl groups of deoxyribose, namely $C^3$ and $C^5$, a polynucleotide column is produced. Notice that such a column has a direction. Thus, if we suppose that it begins with a $C^3$ then it ends with a $C^5$ (Fig. 2.2 and p. 21). This polarity means that if a nucleotide or nucleotide sequence is detached and inverted it cannot then be reinserted into the column (cf. p. 21).

This polynucleotide column has some similarities with the polypeptide chains of proteins (p. 143). Thus the repeating sugar-phosphate sequence

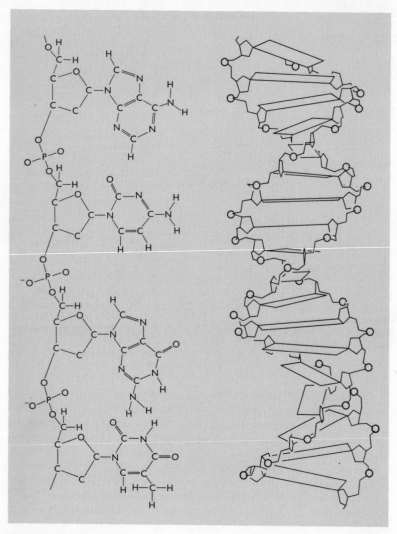

**Fig. 2.2** The polymeric and duplex structure of DNA. The single polynucleotide chain shown contains adenine and cytosine (the two 6-amino bases), and guanine and thymine (the two 6-keto bases) in that order from the 5′ end. The helical duplex formed by hydrogen cross-bonding between the bases of two anti-parallel polynucleotide columns is shown on the right.

gives the polynucleotide chain a regular backbone of

$$C^3—C^4—C^5—O—P—O—C^3—C^4—C^5$$

which can be compared with the equally regular

$$H_2N—C—C\text{-}\text{-}\text{-}N—C—C—N—C—C—N—C—C\text{-}\text{-}\text{-}N—C—COOH$$

sequence produced by peptide linkages. Likewise the bases attached as side chains to the sugar residues of DNA can be compared with the amino acid residues in polypeptides. Variation and specificity in the latter depend on the amino acid sequence and, since it is the only variable feature of the primary structure, we can anticipate that the specificity of DNA likewise depends on the base sequence.

Only in recent years has the amino acid sequence of proteins been determined, and even now the order is established for only a few of the simpler protein molecules. But the over-all amino acid composition of many proteins has been known for some time. In the case of nucleic acids, information on base sequence is meagre and difficult to obtain (but see Fig. 7.7). But determinations of the over-all base composition have revealed a number of interesting and significant features.

In 1949 Chargaff and his associates found that the proportions of the four principal bases were the same in DNA samples taken from different organs of the body and no differences could be detected in this respect between individuals of the same species. Samples from unrelated species, on the other hand, could generally be distinguished on this basis (Table 2.1).

Despite the inter-specific variation, however, certain relationships were

**Table 2.1** Base ratios of some DNAs. N.B. The DNA of $\phi$X174 is single-stranded (p. 23).

| Organism | A/T | G/C | $\dfrac{A+T}{G+C}$ | $\dfrac{\text{6-amino}}{\text{6-keto}}$ |
|---|---|---|---|---|
| 1. Phage $\phi$X174 | 0·75 | 1·31 | 1·35 | 0·76 |
| 2. Phage $T_2$ | 1·00 | 1·09 | 1·87 | 0·97 |
| 3. *E. coli* | 1·09 | 0·99 | 1·00 | 1·05 |
| 4. Yeast | 0·96 | 1·08 | 1·80 | 0·95 |
| 5. Sea urchin | 1·02 | 1·01 | 1·85 | 1·01 |
| 6. Salmon sperm | 1·02 | 1·01 | 1·43 | 1·00 |
| 7. Ox liver | 0·99 | 1·00 | 1·37 | 1·00 |
| 8. Calf thymus | 0·98 | 1·15 | 1·28 | 0·94 |
| 9. Man sperm | 0·98 | 1·03 | 1·67 | 0·97 |

found to hold, within the limits of experimental error, in all the samples analysed. Thus, in every instance:

1. The number of adenine molecules liberated by hydrolysis equalled the number of thymine molecules released.
2. The number of guanine molecules equalled the number of cytosine molecules.

From these it follows that:

3. The number of purine bases equalled the number of pyrimidines.
4. Guanine + thymine equalled adenine + cytosine in amount. In other words, the 6-keto bases equalled in number those bases with an amino group at the 6 position (Fig. 2.1).

The basic construction of a polynucleotide column, as outlined above, does not impose any restriction on the ratio or sequence of bases. But the chemical analyses clearly show that there are restrictions, and proposals regarding the detailed, three-dimensional structure had to take these into account.

While Vicher, Zamenhoff and others were attacking the problem of DNA structure from the chemical point of view, Wilkins and his associates were studying the molecule by X-ray crystallography. They too found uniformity, in that DNA from $T_2$ phage, various bacteria, the trout and the bull all gave the same X-ray pattern.

This pattern indicated that DNA was a helical structure with a diameter of about 20 Å and a pitch of about 34 Å.

Armed with the information from chemical and X-ray diffraction studies and a knowledge of inter-atomic distances and bond angles, Watson and Crick proceeded to construct molecular models of DNA. And, with slight modifications made by Wilkins,[95] the model proposed by Watson and Crick in 1953 has been generally accepted.[91]

The principal features of their model of the paracrystalline structure formed at 90% humidity are as follows:

1. DNA consists of a double, right-handed helix in which two polynucleotide chains are coiled about the same axis and interlocked. This means that they can separate only by untwisting; direct lateral separation is not possible.
2. The bases, as suggested by Astbury and Bell as long ago as 1938, are set in a plane at right angles to the long axis of the molecule.
3. The bases on one chain are joined by hydrogen bonds to those at corresponding levels on the partner chain and there are ten base pairs (and thus, ten sugar-phosphates) to each turn of the helix.
4. Base pairing occurs only between adenine and thymine, on the one

hand, and guanine and cytosine on the other. Thus, the two polynucleotide chains have complementary base sequences.
5. The two chains run in opposite directions (Fig. 2.2 and p. 17).

Clearly, the model accommodates the chemical and X-ray data given
earlier but it has new features also. One of the most important of these
is the restriction the model imposes on the composition of base pairs.
The restrictions imposed are consistent with the observed base ratios
but there were other reasons also for proposing them. The X-ray evidence
indicated a helical structure for lengths equivalent to at least 20 turns of
the helix. If the spiral was regular (i.e. helical), as the model proposed,
the distance between the two backbones, which were held to be on the
outside of the molecule, would be constant over long intervals. Thus, the
space available for the base pairs on the inside of the molecule would also
be constant and limited. A pair of double-ring purines would be too large
to fit the available space. A pair of single-ring pyrimidines, on the other
hand, would be too small, since atoms linked by hydrogen bonds cannot
be more than 3·4 Å apart. The more specific restriction of possible
partners is imposed by the positions of the hydrogen atoms (see p. 39).
Thus, the limitations are imposed by the regularity of the three-
dimensional structure because isolated nucleotides can show a variety
of pairing patterns.

The regularity also means that the orientation of the glycosidic bonds
must be consistent and the two backbones must run in opposite (anti-
parallel) directions. Thus, if from a given physical end of the double
helix the sequence in one backbone is

$$C^3—C^4—C^5—O—P—O—C^3 \text{ etc.,}$$

the sequence in the partner will run

$$C^5—C^4—C^3—O—P—O—C^5 \text{ etc.,}$$

from the same physical end. This means that, although a single polynucleotide column has a 'top' and a 'bottom', the double helix looks the
same when it is turned upside-down. Thus if a single base pair or
sequence of pairs is removed from a terminal or interstitial position and
inverted (rotated through 180°) it can fit into the molecule on reinsertion.
This rearrangement would mean, however, that bases previously located
on one chain would now be transferred to the partner chain (see Fig. 2.3).
Consequently, the model does not impose any restriction with regard to
base content or sequence in any one polynucleotide chain but each chain
must be complementary to its partner in the double helix. The complementary character of the two chains in the duplex DNA provides a basis

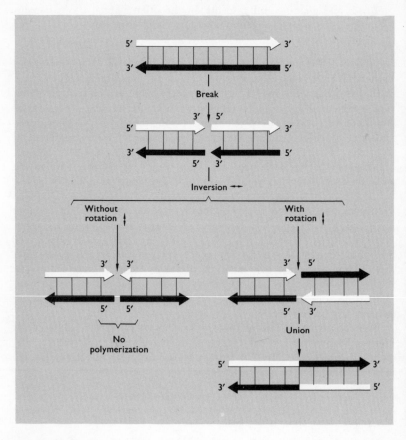

**Fig. 2.3** The inversion of a segment in duplex DNA involves an exchange of material between strands because polymerization links 3′ and 5′ terminals. Without rotation, pairs of 3′-3′ and 5′-5′ ends are brought into opposition and union cannot occur.

for the mechanism of replication suggested by Watson and Crick. They proposed that replication involved the disruption of the hydrogen bonds followed by a rotation and separation of the two chains. Each chain was held to function as a template so that a new complementary partner was polymerized alongside each of them. The net result of this would be the formation of two double helices, identical with each other and with the original helix (Fig. 2.4).

Notice also that the complementary nature of the two chains in the

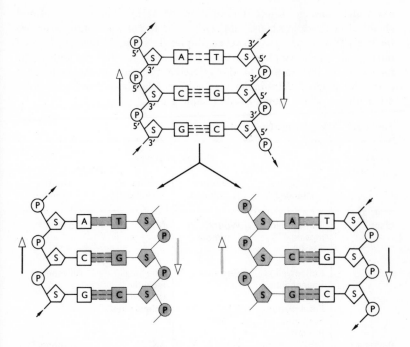

**Fig. 2.4**  The semi-conservative replication of duplex DNA.

duplex and their mutually templating function means that, in a sense, the duplex does not contain any more information than either of the single polynucleotide columns separately. In fact, the base ratios in the DNA of the infective phase of $\phi$X174, one of the smallest viruses, with a particle weight of $6 \cdot 2 \times 10^6$, do not conform with those expected of a double helix composed of two complementary chains (Tables 2.1 and 2.2). Further, the DNA which represents 25% of the particle by weight and has a molecular weight of c. 2 million does not give the X-ray diffraction pattern typical of a double helix. Furthermore, the titration curves indicate that the amino groups, unlike many of the functional groups in the double helix, are not involved in hydrogen bonding and so are available for titration. A similar condition can be achieved by denaturing duplex DNA by heating. This reduces the viscosity of the viscous solutions formed by duplex DNA in the native state and it exposes the functional groups though covalent bonds are not destroyed. Thus, the DNA of $\phi$X174 reacts with formaldehyde while duplex DNA does not, and it is attacked by the phosphodiesterase from the bacterium *Escherichia coli* which does not affect double-stranded DNA. These physical and chemical analyses are consistent with a single polynucleotide structure for the DNA of $\phi$X174. However, duplex DNA molecules are formed during the vegetative phase of this virus (see p. 31). But the genetic information is transmitted to the infecting progeny in the form of a particular single chain. Single-stranded DNA has been described also in other phages which attack *E. coli* (Table 2.2).

**Table 2.2**  The genetic material of viruses

| Host organism | Nucleic acid type | | | |
| --- | --- | --- | --- | --- |
| | RNA | | DNA* | |
| | Single-stranded | Double-stranded | Single-stranded | Double-stranded |
| Plant | Turnip yellow mosaic TMV | Wound tumour | None? | None? |
| Animal | Poliomyelitis Encephalitis Influenza Fowl plague Foot and mouth | Reovirus | One small recently isolated animal virus | All others? |
| Bacteria | MS2 f2 r17 | None? | $\phi$X174 $\phi$R S13 Fd | All others? T2, T5, T7, $\lambda$, P22, SP8 |

* Recent evidence strongly suggests that cauliflower mosaic virus contains DNA.

## The Replication of DNA

Watson and Crick proposed that DNA reproduced in what came to be called a semi-conservative manner because, while the two columns separated from each other, they each retained their longitudinal integrity.

This suggestion was tested in 1958 by Meselson and Stahl who used a heavy isotope of nitrogen to label the DNA and ultracentrifugation to determine the density of the molecule.[66] The basis of the method, developed by Vinograd, is as follows. When strong salt solutions are subjected to high centrifugal force, a continuous gradient of density, increasing from the centre, is set up as a balance is reached between the processes of diffusion and sedimentation. Now, when a substance having a density within that of the gradient is dissolved in the salt solution it will, as it were, find its own level in the gradient. The method is very accurate and the dissolved substance, if homogeneous and heavy, forms a fairly narrow line in the centrifuge chamber. The narrowness of the band is a reflection, but not an accurate measure, of molecular weight. Thus, large molecules, like DNA, which diffuse slowly are effectively concentrated by the forces which oppose diffusion, namely buoyancy at the lower border and the centrifugal force at the upper border of the band. For example, Meselson and Stahl made a culture solution in which all the nitrogen was of the heavy type ($^{15}$N). When bacteria were grown in this solution for a number of cell generations, their DNA became almost fully labelled with the isotope. Meselson and Stahl extracted this DNA and that from a culture grown on normal nitrogen ($^{14}$N) and mixed the two extracts. This mixture was dissolved in a solution of caesium chloride and subjected to ultracentrifugation (44,770 revolutions per minute) until equilibrium was reached (20 hours). The level of the DNA was then determined from photographs showing ultra-violet absorption. The method was found to be sufficiently sensitive to separate the two fractions in the mixture, the buoyant densities of $^{14}$N-DNA and $^{15}$N-DNA from E. coli being 1·709 and 1·725 respectively.

Its suitability having been established, the method was then used to follow the replication of DNA. As before, *Escherichia coli* was grown in the $^{15}$N culture solution for about 14 cell generations. The bacteria were then transferred abruptly to a culture solution in which all the nitrogen was $^{14}$N. Prior to transfer and at intervals afterwards, the cultures were sampled by removing about $4 \times 10^9$ cells, extracting the DNA they contained and subjecting it to the analytical procedure outlined above. Cell counts showed that, under the conditions used, the cell generation time was about 50 minutes and the growth of the cultures was exponential.

The results of the experiment were beautiful in their simplicity (Fig. 2.5). The DNA sample which was obtained one cell generation after transfer to $^{14}$N formed a single band in the density gradient, thus indicating homogeneity. At equilibrium this band was exactly ($\pm 2\%$) halfway between the two bands formed by $^{15}$N-DNA and $^{14}$N-DNA.

The DNA in the sample taken two cell generations after transfer to the $^{14}$N solution formed two bands of equal intensity. One of these coincided with the $^{14}$N band while the other, like the single band of the previous generation, was mid-way between the $^{14}$N and $^{15}$N bands. These same two bands (light and intermediate) were given by the DNA of all subsequent generations. But the relative intensity of the intermediate band decreased progressively while that of the other increased. However, there did not appear to be any over-all loss of the 'intermediate molecules' though by dilution they seemed to disappear. In other words, the heavy chain of the original hybrid, $^{15}$N-N$^{14}$ molecule retained its integrity.

It was further found that if the DNA which formed the intermediate

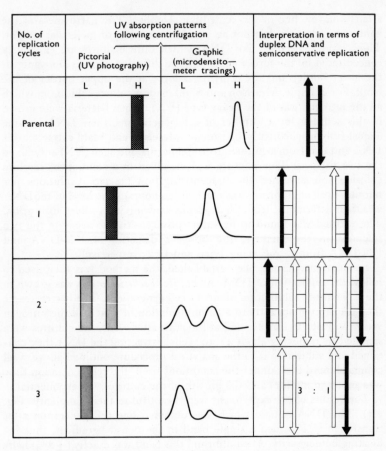

**Fig. 2.5** The pattern of incorporation of [14]N into uniformly [15]N-labelled DNA over 3 replication cycles in *Escherichia coli*. The amount of DNA at various levels in the centrifuge tube is indicated by the intensity of the image in UV photographs and by the tracings obtained with microdensitometry. The density of the CsCl gradient increases to the right. In terms of the hypothesis of semi-conservative replication for duplex DNA, the Light (L), Intermediate (I) and Heavy (H) bands contain homoduplex [14]N-N[14], heteroduplex [14]N-N[15], and homoduplex [15]N-N[15] molecules, respectively. The figure is diagrammatic and somewhat idealized.

band was heated for half an hour at 100°C and then centrifuged, two equal bands were obtained. The positions of these bands corresponded with those of the two bands obtained separately when pure [14]N-DNA and pure [15]N-DNA were given the same treatment.

These results clearly indicate a semi-conservative pattern of replication and, if the gene string (genoneme) of *E. coli* is mononemic (i.e. only one duplex thick) prior to replication, the columns which separate and maintain their longitudinal integrity must be single polynucleotide columns. And the evidence is that they are mononemic (see below).

The general consensus, therefore, is that DNA replication is semi-conservative; the two chains separate from one another, each acts as a template in the synthesis of a new complementary partner, and each single polynucleotide column of the original double helix retains its integrity.

But how long are these molecules? How much DNA is there in a bacterium and its viruses, and is it all in one piece? Conceivably the DNA in a chromosome, for example, could consist of a number of separate molecules joined end to end by labile linkages and while each separate molecule might replicate in a semi-conservative manner the whole chromosome might not. Indeed, perhaps this applies in bacteria also for it could be argued that labile linkages are broken during extraction and centrifugation. In fact it is known that extracted DNA is very susceptible to degradation by shearing forces. This contributed to the comparatively low molecular weights determined for DNA in some of the earlier work and for the heterogeneity of weights within samples.

However, sedimentation studies on carefully extracted DNA of various phage particles (e.g. $T_2$, $T_5$ and $\lambda$) have shown that the molecular weights correspond with the total DNA of the individual phage concerned. In other words, the DNA is in one piece. What is more, auto-radiographic and electron microscope studies have shown that in many viruses, like $T_2$, the double helix apparently forms a circle. Indeed this appears to be true also in the vegetative cells of *E. coli* where the DNA ring has a circumference of over a millimetre (but see pp. 67 and 109). This circular structure is in keeping with the genetic evidence but, in view of the previously known linear linkage maps of higher organisms, the discovery came as something of a surprise.

Of course, these circles need not conform completely to the Watson–Crick model but departures from it are not likely to be extensive. Thus, the ratio of mass to length in the DNA of $\lambda$-phage is $2 \times 10^6$ molecular weight units/$\mu$ and, allowing 330 as an average molecular weight for a nucleotide pair, this ratio corresponds with that of the paracrystalline form of DNA. It also indicates a mononemic structure.

Clearly, the Meselson and Stahl experiment is relevant to only a particular aspect of DNA replication, for it examines not the sequence but the consequence of the process. Information concerning the sequence comes from the autoradiographic studies of Cairns on *E. coli*.[16] The basis of the autoradiographic method is simple.

First the test material is supplied with a suitable radioactive material which then becomes incorporated into the critical molecule or structure under investigation. In the case of DNA, labelled thymidine ($^3$H-TdR) is usually employed because this is specifically incorporated. When non-specific precursors are used, extraneous material which takes up the label can often be selectively broken down by enzymes. The material is then prepared (e.g. sectioned), covered with a photographic emulsion and kept in the dark. The particles emitted by radioactive decay 'expose' the emulsion which is then developed in the usual way. Under the microscope the track of the particle will appear as silver. If radioactive materials which emit particles of low penetration are used (e.g. tritium emits β-particles of this kind), the site of a silver grain corresponds very closely with that of the atom from which the particle was emitted.

In an exponentially-growing culture of bacteria, DNA is synthesized more or less continuously and under the conditions used by Cairns for *E. coli* the generation time was about 30 minutes. He grew this bacterium in the presence of tritiated thymidine for one hour, removed the DNA and subjected it to autoradiographic analysis. He found some 'nuclei' which had started but not completed the second replication cycle in the presence of radioactive precursor. Autoradiographs of such nuclei had the appearance shown diagrammatically in Fig. 2.6.

Clearly the split region is that which has completed the second cycle in the presence of the label. Double splits were never seen. This shows that replication is initiated at one point ('replicator' site) from which it

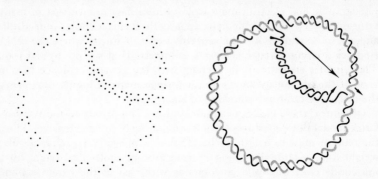

**Fig. 2.6**  The replication of the circular genoneme of *E. coli*. An autoradiogram of a nucleus during the second cycle of DNA replication in the presence of $^3$H-thymidine is shown diagrammatically on the left. One arc of the split region shows twice the grain density of other regions which are equally labelled. An interpretation in terms of unidirectional semi-conservative replication from a fixed point is shown on the right. Note that the two strands of the original duplex cannot unwind unless some breakage occurs or at least one region has special rotational properties.

then proceeds. In such nuclei one arc of the split was always found to be twice as heavily labelled as the other arc which showed the same density of labelling as the unsplit, unreplicated region. In other words, at no level in the split region, not even near the Y junctions between the single and double regions, did the two arcs of the split region show single versus single labelling, no labelling versus single labelling, or no labelling versus double labelling.

These observations clearly support the semi-conservative nature of DNA replication. They also indicate that replication proceeds more or less simultaneously at corresponding levels in the two columns produced by the separation of the chains in the original duplex. In other words, replication starts at the same point in the two chains and progresses in the same physical direction in both of them.

However, these observations do not show whether the point of replication initiation is constant as between nuclei and cells, nor do they indicate whether replication proceeds in one or both physical directions from the replicator. Information on these questions was, however, obtained by Cairns from the autoradiograms of 'chromosomes' which had incorporated label only during the later stages of the first replication cycle and were in the process of undergoing the second cycle in the presence of label. The label patterns shown by these 'chromosomes' were consistent with the conclusion that the point of replication initiation is fixed and that synthesis proceeds in one physical direction from this point. The studies of Yoshikawa and Sueoka have a bearing on these questions also.[98] They argued that, for the most part, the DNA in the cells of a bacterial culture during the stationary phase of growth will not be in the process of replication; they will contain complete genonemes rather than partially replicated ones. Thus, all parts of the genoneme will be equally represented numerically. During exponential growth, on the other hand, DNA synthesis proceeds almost continuously and partially replicated genonemes are common. Now, if replication is initiated at the same point in all cases, then regions near the origin will be replicated first and they will be represented about twice as often as those near the terminus of replication. Further, in an asynchronous culture, intervening regions will be represented in proportion to their distance from the replicator, assuming that replication proceeds at uniform rate (Fig. 2.7).

Now the success with which a given character can be transformed (see p. 7) varies with the character concerned (i.e. with the nature of the mutation involved) but it clearly depends also on the amount of corresponding DNA in the transforming principle. Therefore, if DNA extracted from asynchronous, log-phase cultures is used to effect transformation, and the success of the venture is compared with that obtained with DNA from the stationary phase, relative differences should be observed for different regions depending on their location in relation to the replicator.

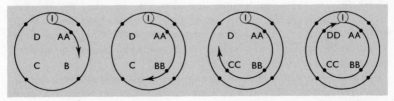

**Fig. 2.7**  Diagrammatic representation of four successive stages in the replication of a circular genoneme. Replication is assumed to proceed in a fixed direction from a fixed point (I). Single lines are used to represent double helices. In asynchronously growing populations, all phases of replication will be represented and, where continuous replication in the log-phase is oriented and progressive, regions will be represented in proportion to their 'clockwise' direction distance from the Initiator. Thus, in the diagram, A = 8, B = 7, C = 6 and D = 5. Differential representation will be paralleled by different rates of transformation when DNA from asynchronous cultures is used. This is the rationale behind the experiment of Yoshikawa and Sueoka (see p. 99).

Yoshikawa and Sueoka studied the transformation of 10 different traits in *Bacillus subtilis* and found that these could be arranged in a linear order at distances apart proportional to the relative success of the transformation achieved by log-phase DNA. Further, the relative success obtained for the character placed at one end of this map was about twice as great as that for the character at the other end.

If, now, this map can be shown to correspond with the linkage map obtained by other means (see p. 86), the unidirectional synthesis of DNA from a fixed initiator is established. Unfortunately, detailed maps of this kind are not available for *B. subtilis*. It was shown, however, that certain regions which appeared close together on the transformation map also showed joint transformation more often than expected on the basis of the coincidence of independent events (see p. 91). It appears, therefore, that in organisms with a simple genoneme organization, DNA replication is initiated at a fixed point from which it then proceeds in one physical direction (see also p. 167).

Now, the two polynucleotide columns in duplex DNA are antiparallel. Consequently, if synthesis proceeds simultaneously at exactly corresponding levels in the new chains, so that they grow continuously in strictly the same physical direction, nucleotides must be added to the 3' end of one growing chain and to the 5' end of the other. In other words, growth in the same physical direction implies growth in opposite chemical directions.

However, the resolving power of the autoradiographic method, and especially that of genetic techniques using widely-spaced markers, is not sufficiently high to warrant this conclusion. Thus, although the general pattern of replication is progressive and unidirectional from a fixed point, it cannot be concluded on the above evidence that DNA can replicate in both chemical directions. In fact, it is likely that, in one of the chains, at least, discontinuous DNA synthesis occurs within segments about one or two thousand bases long. This view is consistent with the evidence from *in vitro* DNA synthesis. Thus, DNA of the same base composition as highly polymerized, single- or double-stranded, primer can be synthesized *in vitro* with the aid of an enzyme obtained from *E. coli*. This polymerase acts in one direction only, adding 5′ nucleotides to the 3′ nucleoside end as opposed to the phosphate end of the growing chain.

The experiments described above relate to the organization of DNA replication in organisms with a simple genoneme constitution.[62] Comparable studies have been performed in organisms with a true chromosome structure in which DNA is complexed with protein. Thus, density gradient studies of the type performed by Meselson and Stahl on *E. coli* were subsequently undertaken in *Chlamydomonas reinhardi* and human cells with similar results. In the latter investigation, 5-bromo-uracil was used as a heavy label. Autoradiographic studies on chromosome replication in higher plants, on the other hand, were actually performed prior to those on bacteria and phage. In general, semi-conservative replication appears to be the rule (if not an invariable one) at this level also. But the evidence does not indicate a linear progression of synthesis extending the whole length of a chromosome or even a chromosome arm.

The replication of single-stranded DNA has been studied in $\phi$X174. Under *in vitro* conditions, the single-stranded DNA of this phage can function as a template in the production of duplex molecules, both strands of which can serve as templates in subsequent replication cycles. Whether both strands of such a duplex are equally copied under natural conditions is doubtful (see p. 37). It is clear, however, that a two-stranded, replicative form (RF) of DNA is formed *in vivo*. This was demonstrated by infecting bacterial host cells grown on a medium containing $^{31}P$ and $^{14}N$ with doubly $^{32}P^{15}N$-labelled phage. After density gradient centrifugation, the DNA which was extracted during the period of viral DNA synthesis separated as two bands both of which showed radioactivity derived from the original DNA. However, while the heavier band showed the density expected of single-stranded, $^{15}N$-labelled DNA, the density of the light band was that expected for a hybrid $^{15}N$-$N^{14}$ double helix. The base composition of this duplex was consistent with a complementary structure and it showed the expected behaviour on heating. In this case also there is some evidence that the double helix forms a circle.

## RIBONUCLEIC ACID

### The structure of RNA

Techniques comparable with those used in the study of DNA have shown that RNA is very similar to DNA in its general composition and primary structure. Thus, RNA consists essentially of chains with a repeating backbone of phosphate and sugar units to which nitrogenous bases are attached. The similarity between single RNA and DNA poly-nucleotide columns is such that a duplex composed of both can be produced so long as the base sequences permit cross-bonding (p. 164). The differences between these two nucleic acids are of three main kinds.

First, the D-ribose sugar unit of RNA differs from the 2-deoxy-D-ribose of DNA in having a total of four rather than three hydroxyl groups. The extra OH, as the names indicate, is at the $C^2$ position. Its presence introduces the prospect of polymerization by means of $C^2$–$C^5$ phospho-diester linkages but it would appear that, as in DNA, so in RNA, $C^3$–$C^5$ internucleotide linkages are actually involved in polynucleotide formation.

Second, both DNA and RNA contain four principal bases but only three of these, adenine, guanine and cytosine, are common to the two nucleic acids. The RNA counterpart of thymine is uracil, the former being the 5-methyl derivative of the latter. Like DNA, RNA too may contain unusual or minor bases in small quantities. Indeed, even thymine has been found in some RNA preparations. The glycosidic linkages between sugar and base in RNA are the same as those in DNA (Fig. 2.8).

The third difference relates to the secondary and tertiary structures. RNA is not as well understood as DNA in this respect and difficulties

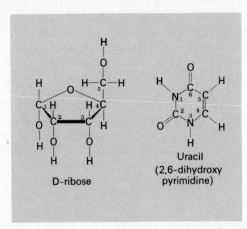

**Fig. 2.8**   The sugar and base components which distinguish RNA from DNA.

are introduced by the functional and structural heterogeneity of RNA. Thus, four main types of RNA can be distinguished which can be considered in two classes:

1. GENETIC RNA    In most plant viruses, RNA rather than DNA provides the chemical basis of heredity (Table 2.3). RNA serves this function also in certain bacteriophages (e.g. f2, MS5 and r17)[100] which attack *E. coli* and in some animal viruses (e.g. poliomyelitis virus and the polyhedral cytoplasmic viruses of insects). The RNA may represent as little as 1% of the virus particle (e.g. influenza virus) or as much as 40% (tobacco ring spot).

2. NON-GENETIC RNA    While RNA can serve genetic functions, it does not appear to do so in organisms which contain DNA. Non-genetic RNA is specified by the genetic material (see p. 159) and is itself heterogeneous, three principal genera being distinguished:

1. Ribosomal RNA (rRNA). Ribosomes are nucleoprotein particles about 20 m$\mu$ in diameter, composed mainly of basic protein and RNA to the extent of 40–60% by weight.[69] In *E. coli* two types of ribosomal sub-unit are found, the sedimentation coefficients of which are 50S and 30S (S = Svedberg units). One of each of these sub-units may aggregate to give a 70S particle which can pair to give 100S complexes. Ribosomal RNA represents about 80% of the total cell RNA.
2. Soluble RNA (sRNA). This genus is the next most abundant and contributes about 10–15% to the total RNA. The soluble RNA consists of small molecules which are uniform in regard to size. Their sedimentation coefficient is 4S and their molecular weight is in the region of 25,000. Unusual bases are prominent among these small molecules (Table 2.4). This fraction includes the 'adaptor' or 'transfer' RNA (tRNA) molecules which act as amino acid acceptors during protein synthesis. Each kind of transfer RNA is believed to be specific for a particular amino acid, though a given amino acid may be accepted by more than one kind of transfer RNA molecule (see p. 162). However, it seems that all the RNA molecules in the soluble fraction cannot or do not serve this adaptor function.
3. Messenger RNA (mRNA). About 5–10% of the RNA in a cell exists in a short-lived form which functions as a specific template in protein synthesis. The molecular weights reported for this fraction vary widely as do the S coefficients but molecular weights between half and two million are usual. It would seem, however, that the genetic RNA of viruses can function directly as messenger.

At this stage only the nucleic acid of RNA viruses will be considered in any detail for, while it can function as a messenger, it appears to be

**Table 2.3**  The chemistry of viruses.

| Host | Virus | NA type | Virion size (mμ) | Virion M. Wt. | % | NA M. Wt. | G | C | A | T(U) |
|---|---|---|---|---|---|---|---|---|---|---|
| Plant | TMV | $R^S$ | 15 × 300 | 40 × 10⁶ | 5·0 | 2 × 10⁶ | 25·4 | 18·5 | 29·8 | 26·3 |
| | Turnip yellow mosaic | $R^S$ | 70 | 5 × 10⁶ | 35·0 | 2·3 × 10⁶ | 17·2 | 38·0 | 22·6 | 22·2 |
| | Wound tumour | $R^D$ | | | | (10 × 10⁶) | 18·6 | 19·1 | 31·1 | 31·3 |
| Animal | Influenza | $R^S$ | | 280 × 10⁶ | | 1 × 10⁶ | 18·2 | 24·0 | 22·25 | 30·33 |
| | Foot and mouth | $R^S$ | | | | | 24·0 | 28·0 | 26·0 | 22·0 |
| | Rous sarcoma | $R^S$ | 50 | 6·8 × 10⁶ | | 9·5 × 10⁶ | 28·3 | 24·2 | 25·1 | 22·4 |
| | Poliomyelitis | $R^S$ | 70 | >70 × 10⁶ | 30·0 | 2·2 × 10⁶ | 24·0 | 22·0 | 25·4 | 25·4 |
| | Reovirus | $R^D$ | | 40 × 10⁶ | 14·6 | >10·2 × 10⁶ | 19·3 | 20·5 | 29·7 | 30·5 |
| | Polyoma | $D^D$ | 150 | ~10⁹ | | 5·3 × 10⁶ | 24·0 | 24·0 | 26·0 | 26·0 |
| | Herpes | $D^D$ | | 2000 × 10⁶ | | 60–80 × 10⁶ | 38·0 | 35·0 | 14·5 | 13·1 |
| | Vaccinia | $D^D$ | | | | 150 × 10⁶ | 18·0 | 19·0 | 31·5 | 31·5 |
| Bacteria | MS2 | $R^S$ | | 3·6 × 10⁶ | 31·0 | 1 × 10⁶ | | | | |
| | φX 174 | $D^S$ | | 6·2 × 10⁶ | | 1·7 × 10⁶ | 23·0 | 19·0 | 25·0 | 33·0 |
| | T₂ | $D^D$ | | | | 1·2–1·6 × 10⁸ | 18·2 | 16·8* | 32·5 | 32·5 |

\* N.B. In $T_2$, cytosine is replaced by 5-hydroxymethylcytosine
S = Single-stranded
D = Double-stranded

**Table 2.4**   Unusual bases in sRNA. (After Jukes, T. H. (1966) *Molecules and Evolution*, Columbia University Press, New York and London.)

| Base | Source |
|------|--------|
| 1. 1-Methyladenine | Pig liver |
| 2. 2-Methyladenine | S-180 ascites |
| 3. 6-Methyladenine | Rat liver |
| 4. 1-Methylguanine | Rat and rabbit liver |
| 5. 2-Methylguanine | Mouse adenosarcoma |
| 6. 7-Methylguanine | Pig liver |
| 7. 3-Methylcytidine | Yeast |
| 8. 5-Methylcytidine | Rat liver |
| 9. Inosine | Yeast |
| 10. 2-thiouridine | *E. coli* |

alone among the various genera of RNA in serving a genetic role. The other forms of RNA will be discussed later in relation to their metabolic activities (see p. 158).

*Viral RNA*

DNA has a well defined secondary and tertiary structure and it usually occurs in the form of a rigid and stable double helix composed of two complementary strands. Strand separation does, of course, occur during replication and transcription (see p. 164) but otherwise single-stranded DNA is not common.

The reverse is true of RNA. Thus a double helical structure has been claimed for the RNA of some viruses, e.g. reovirus and wound tumour virus, and the complementary nature of the base ratios supports this view (Table 2.3). Further, a temporary duplex is produced during the replication of viral RNA. But the usual form of genetic RNA in the infective phase of viruses is that of a single polynucleotide chain. This single-stranded state is reflected in the asymmetry of the base ratios of potential pairing partners (Table 2.3).

The tobacco mosaic virus (TMV) has many claims to fame. It was the first virus to be discovered (1892), the first to be isolated (1935), the first to be fractionated into its RNA and protein components and the first to be reconstituted from these fractions (see p. 12). The infective phase of this virus consists of protein units which describe a hollow cylinder 3000 Å long with an external diameter of 150–180 Å and an empty axial core believed to be 40 Å in diameter. The protein units describe a regular spiral with a pitch of 23 Å and 49 of these units occupy three turns of the helix.

The RNA consists of a single polynucleotide column. This is threaded through the protein units (Fig. 1.2) following a helix 80 Å in diameter and having the same gentle pitch as the protein sub-units to which it is bonded by three phosphate groups per sub-unit. Unbroken RNA derived from the virus retains its infectivity and these unbroken lengths have a sedimentation value of 30S which corresponds with a molecular weight of $2 \times 10^6$. These values support the evidence of electron microscopy and X-ray diffraction in showing that each particle has only one molecule of RNA and, on the evidence of its viscosity, light-scattering properties and the kinetics of its inactivation by heat or enzymes, this molecule is, as the base ratios suggest, single. Thus, the genetic material of TMV consists of about 6400 nucleotide units.

Although the backbone of a single polynucleotide column contains rotatable bonds—rather like a series of ball and socket joints—the RNA in the intact TMV is constrained into a regular helix by its protecting protein coat and this configuration does not allow inter-base connections. However, connections of this kind may occur in simpler viruses where the RNA may be tightly packed into a central core. In fact, the presence of hydrogen cross-bonding is indicated by the increase in UV absorption which follows the alkaline hydrolysis of simple, spherical, plant viruses like turnip yellow mosaic. Likewise there are indications of a high degree of H-bonding in the RNA of bacteriophage MS2. This particle has a weight of about $3 \cdot 6 \times 10^6$ and contains 30% RNA.

It must be remembered, however, that while we tend to think of the virus in terms of its discrete, countable, and often crystallizable, infective phase, the biologically active virus is in nothing like this condition. In regard to its genetic behaviour and metabolic function, it is not the state of the nucleic acid in the infective phase that matters but its condition in the host cell. And the configurations adopted by RNA in solution may give a better indication of its state in the cell. In passing let us also note that we often tend to approach chromosomes in much the same way, the compact, countable chromosomes of division being more tangible than the diffuse material of the optically empty metabolic nucleus.

## The replication of RNA

It has been suggested that ribosomal and soluble RNA, though initially specified by DNA, may show some self-replication. The available evidence, however, is against this view and the only known instance of RNA-dependent RNA synthesis is the replication of the genetic material of RNA viruses.

Several models have been proposed for this process. For example, it was once suggested that the viral RNA first specified a complementary DNA polynucleotide which subsequently directed the synthesis of genetic, viral RNA. Various lines of evidence are against this view. For example, the f2 RNA phage of *E. coli* has been shown to replicate in a normal manner even when DNA synthesis is inhibited. Again, antibiotics like actinomycin D which complex with DNA, and so inhibit both its replication and, especially, its transcription, have no effect on the replication of the RNA phage MS2. In addition the RNA of this virus will not hybridize with DNA from normal or infected host cells. Further, mammalian cells infected with EMC virus do not show increased activity of either DNA polymerase or DNA-primed RNA polymerase.

In fact, the available evidence strongly suggests that the synthesis of viral RNA depends on the RNA itself and a number of features of RNA virus infection have now been elucidated. For example, it has been shown that when the RNA from phage f2 is isolated and introduced into an *in vitro* system capable of supporting protein, but not RNA, synthesis, polypeptides characteristic of the viral coat protein are produced. This indicates that the genetic RNA can function directly as a messenger without further transcription.

It has also been shown that a few minutes after *E. coli* is infected with $^{32}$P-labelled MS2 virus, an RNA complex can be isolated which, on the basis of resistance to ribonuclease attack, density and 'melting temperature', can be identified as a double-stranded RNA molecule. The radioactivity of this duplex is confined to the so-called 'plus' strand derived from the original infective phase of the phage. What is more, an RNA-dependent RNA polymerase (synthetase or replicase) also begins to appear soon after infection and reaches a maximum after about 40 minutes. This enzyme can be isolated together with the duplex replicating form of the RNA and incubated in the presence of labelled triphosphate derivatives of the RNA bases. Of the radioactivity which becomes incorporated in the acid insoluble products, about a half is found in double-stranded RNA molecules and, in these, as much as 85% of the activity is contributed by 'plus' strands. This indicates the preferential production of RNA similar to that in the infective phase.[93]

On the basis of results such as these, Ochoa has proposed the following pattern of events following RNA-virus infection:

1. The viral RNA functions directly as a messenger RNA which, in association with the ribosomal apparatus of the host, directs the synthesis of both the RNA polymerase and the proteins of the viral coat (see p. 158).
2. With the mediation of the RNA polymerase, and on the standard base-pairing principles, the viral RNA serves as a template in the

synthesis of a complementary 'minus' chain and a double helix is produced. This duplex can be compared with the double, replicating form of the single-stranded DNA phage φX174 (p. 31).

3. In the same way, the newly-formed minus strands produce a series of new plus strands, each new plus strand replacing the previously synthesized one. Thus, a large number of plus strands, which correspond to the RNA of the original infective phase, is produced. These and the protein units of the coat then aggregate to reproduce new infective phage particles.

This scheme was formulated originally in relation to the MS2 phage but it appears to hold in other cases also.[64] Thus, RNA-dependent RNA polymerase has been found in mammalian cells infected with mengo and polio virus and in virus-infected plant cells. In fact, TMV gets another first in this connection for the earliest demonstration of RNA-dependent RNA polymerase was provided by Reddi in 1961 when he showed that RNA synthesis in spinach extracts was stimulated by RNA from this virus. Further, the duplex replicating form of viral RNA has been found both in TMV and EMC (encephalomyocarditis) virus.

Some caution must, however, be expressed for it is frequently unwise to equate *in vivo* behaviour with that found *in vitro*. For example, *in vitro* DNA-dependent RNA polymerase appears to copy both strands of the DNA double helix. However, there is good evidence to show that only one particular strand is transcribed *in vivo* (see p. 164). Similarly, differences are known also in relation to *in vivo* and *in vitro* systems of DNA-dependent DNA polymerase.

# 3

# *The Chemical Basis of Mutation*

Genetic material must be able to contain and maintain information which can be duplicated and transmitted during heredity and transcribed and translated during development. Clearly, this information must be coded in terms of the four-letter alphabet of the nucleic acid bases, because the base sequence is the only variable feature in the primary structure of both DNA and RNA.

Thus, the stability of heredity must rest ultimately on the accuracy with which the base sequence is conserved and copied.[50] But the stability required of genetic material is a limited one and, in the long run, mutability is a no less important property. Genetic changes may involve alterations in the absolute amounts, relative frequencies or sequence of the bases. Thus, while heredity depends on the accuracy of replication, heritable variation depends on its inaccuracies.

## BASE ANALOGUES AND TEMPLATE SPECIFICITY

Certain reservations have been expressed regarding the sufficiency of complementary base pairing as the sole controlling factor in determining the base composition of DNA following replication. For example, the regularity of the helix shows that, aside from infrequent mistakes, the possibility of pairing is restricted to the complementary pairs adenine-thymine (A-T) and guanine-cytosine (G-C). But, clearly, since the final structure can hardly be invoked as a selective factor in its own production, this fact does not account for the manner in which other pairings are excluded. Of course, incorporation requires that two conditions be fulfilled. First, the precursor must pair with the template and, second, it must form sugar-phosphate links with adjacent members of the new poly-nucleotide column. If the latter is not achieved, the precursor will eventually leave the template. Certain pairings are prohibited on this basis but associations other than those between A-T and G-C are com-patible with both these conditions though the spiral resulting from their inclusion would not be a regular helix.

The question of whether the hydrogen bond relationships are sufficiently specific wholly to determine the double structure is raised also by the behaviour of certain natural and artificial base analogues.

Thus, while A, T, G and C predominate in the DNA of most species, some of these bases are replaced completely or in part by certain deriva-tives in many cases. For example, in wheat about a quarter of the cytosine fraction exists as the 5-methyl derivative, the joint molar concentrations of both 6-amino pyrimidines equalling that of guanine, the 6-keto purine. Clearly, both cytosine and 5-methyl cytosine have, as expected, similar base-pairing properties. 5-methyl cytosine occurs also in other grasses and, in lesser amounts, in various mammals as well (Table 3.1). Now, if

**Table 3.1** Base composition in organisms whose DNA includes 5-methyl cytosine. (Data of Doskocil, J. and Sorm, F. (1962) *Biochim. biophys. Acta*, **55**, 953–9.)

| Base composition (%) | Material Calf thymus | Rat spleen | Mouse liver | Wheat germ |
|---|---|---|---|---|
| Adenine | 28·3 | 28·5 | 28·2 | 27·3 |
| Thymine | 28·6 | 28·7 | 27·9 | 27·6 |
| A/T | 0·99 | 0·99 | 1·01 | 0·99 |
| Guanine | 21·5 | 21·3 | 21·6 | 22·6 |
| Cytosine | 20·1 | 20·2 | 21·0 | 15·8 |
| 5-methyl cytosine | 1·5 | 1·3 | 1·3 | 6·7 |
| $\dfrac{G}{C+M}$ | 0·99 | 0·99 | 0·97 | 1·00 |

we consider only base-pairing and polymerizing relations, there is no obvious stereochemical basis for discriminating between cytosine and its 5-methyl derivative. And if the methyl derivative substitutes for cytosine at random in the DNA molecules that contain it, the question of whether replication is based solely on polynucleotide complementarity is not raised. However, an examination of dinucleotides obtained from partial hydrolysates of DNA from calf thymus suggests that there are considerable differences between the distribution of cytosine and 5-methyl cytosine. This seems to hold also in the case of wheat. Thus it would appear that in mammals this methylated base (M) is found only in the sequence MpG. This location predominates in wheat also but, in addition, the MpT sequence is not uncommon and small amounts of both MpC and MpM are found. The greater variation in wheat probably depends on the greater over-all level of substitution as compared with mammals.

Similarly, in the T-even phages of *Escherichia coli*, cytosine is replaced completely by 5-hydroxymethyl cytosine, about three-quarters of which occurs conjugated with glucose. Here too it appears that the distributions of 5-glucosyl hydroxymethyl cytosine and of 5-hydroxymethyl cytosine are not comparable.

Other naturally occurring base analogues include 5-hydroxymethyl uracil which has been found in certain viruses and 6-methyl purine, an adenine analogue, which occurs in some bacteria.

As might be expected, certain substituted bases, which do not normally occur in DNA or in a particular DNA, can be incorporated under experimental conditions. For example, the 5-hydroxymethyl cytosine found in T-even phages does not occur in the DNA of the host bacterium *E. coli*. However, a polymerase obtained from *E. coli* will mediate in the *in vitro* incorporation of this base in the presence of primer DNA obtained from a variety of sources, including $\phi$X174 and calf thymus as well as *E. coli* itself. In fact, again as expected on the principle of complementarity, many analogues can substitute for the normal base both *in vitro* and *in vivo*. But this is not true for all analogues. For example, uracil has similar properties to thymine, its 5-methyl derivative, which, in fact, it replaces in RNA. It can also replace it in DNA but only *in vitro*, for although uracil occurs abundantly in cells, it does not occur in natural DNA. In this case, however, stereochemical considerations are not involved but rather the absence of the appropriate enzyme which would mediate in its incorporation or else the specific hydrolysis of the relevant deoxyribonucleotide phosphate. The absence of cytosine from the DNA of T-even phages may be similarly explained.

On the other hand, certain *in vitro* systems of DNA synthesis have been shown to incorporate analogues like 5-bromo- and 5-iodo-uracil in place of thymine and 5-bromo- and 5-methyl-cytosine, in place of cytosine, but although these systems accept 5-fluorocytosine and 5-fluoro-uracil, N-methyl 5-fluorocytosine is rejected. Likewise, although hypoxanthine (6-hydroxy purine) is incorporated by these systems in place of guanine, xanthine (2,6-dihydroxy purine) is not.

The pattern of base analogue incorporation shows three further features of significance in regard to the self-sufficiency of the complementarity principle. First, the rate and extent of analogue incorporation vary from nought to about twice those of the normal base. Second, when base analogues are incorporated in competition with their normal equivalents, the sites of analogue substitution do not appear to be randomly distributed in relation to the distribution of the normal base in the native DNA duplex. In other words, the analogue, despite its similar base-pairing properties, is incorporated at preferred locations (Table 3.2). Third, the presence of a base analogue at a particular site appears to interfere with the templating function of normal bases at neighbouring sites.

**Table 3.2**  Distortion by 5-bromo uracil of the base sequence in the DNA of *E. coli*. SM = synthetic medium. The fragments analysed in the lower part of the table were obtained following hydrolysis with 0·1M $H_2SO_4$ at 100° C for 30 min. (Data of Shapiro, H. S. and Chargaff, E. (1960) *Nature, Lond.*, **188**, 62–3.)

| Units | Item | Base composition of DNA | |
|---|---|---|---|
| | | SM + T | SM + T + BU |
| Moles per cent | BU | — | 9·0 |
| | T | 24·8 | 16·6 |
| | C | 25·5 | 25·1 |
| | G | 25·0 | 24·3 |
| | A | 24·8 | 25·3 |
| Molar ratios | $\dfrac{\text{6-amino}}{\text{6-keto}}$ | 1·01 | 1·01 |
| | $\dfrac{\text{A}}{\text{T or (T + BU)}}$ | 1·00 | 1·00 |
| Moles/ 100 g atoms of total DNA P | pBUp | — | 2·96 |
| | pTp | 3·82 | 3·01 |
| | pCp | 1·78 | 3·60 |
| | pCpCp | 0·37 | 0·58 |
| | pCpTp | 1·13 | 1·38 |
| | Frequency of solitary and coupled pyrimidine nucleotides | | |

It must be admitted that some of these results are not consistent with the other experimental evidence but results such as these have been used to support the suggestion that forces and specificities other than those of hydrogen bonding between complementary base pairs are involved in distinguishing between structural analogues. Of course, in a sense, other forces and specificities must be involved because DNA synthesis ultimately depends on a variety of metabolic pathways. Further, the net interaction energy which follows the change from solvent-base to base-base hydrogen bonding is too small to drive the templating reaction and it is unlikely that the initial base-base attractions depend on contacts of such high stereospecificity as those of hydrogen bonding. But the principal point at issue is whether the template is selective in ways other than those indicated by hydrogen bond complementarity and the prospects of phosphate-ester polymerization. In this connection the following must be borne in mind in regard to the above evidence.

First, in cases like the T-even phages where the native DNA contains two forms of the 6-amino pyrimidine, it is not clear that the analogue as such is incorporated at the time of replication. The substitution at the 5-position may occur *after* incorporation. Under these circumstances one would clearly expect the nature of the adjacent bases to affect the incidence and location of substitution. This argument could perhaps be extended to those animals and grasses which have both cytosine and 5-methyl cytosine as normal constituents of their DNA.

Second, it is obvious that DNA replication involves more than a simple template and randomly distributed precursors. Differentials between analogues in regard to the extent and rate of incorporation could be determined at any of the many steps leading to the final synthesis. Thus, these too need not offend the principle of complementarity because they could depend on enzyme specificities of different kinds and not on the configuration of the template. This argument has been outlined above in relation to the absence of cytosine from the DNA of T-even phages and the general exclusion of uracil, although both bases are common cell constituents.

Third, the claim that artificial analogues, while they quantitatively replace their normal base equivalents, are incorporated at preferential locations presents a more difficult problem. Of course, it must be recognized that *in vitro* systems may not be wholly comparable to those *in vivo*. But as shown in Table 3.2, severe distortions of the base sequence have been claimed following the incorporation of analogues *in vivo*. The incorporation of analogues often leads to increased mutation rate and to death. Their roles in these connections are not fully understood and it is to these problems that we now turn.

## CHEMICAL MUTAGENESIS

The specific composition of the purine-pyrimidine base pairs in duplex DNA is determined principally by the positions of the hydrogen atoms responsible for cross-bonding and the prospects of polymerization. Normally, pairing is between A and T and between G and C. However, these bases can exist in alternative, valency-consistent states owing to re-arrangements (tautomeric shifts) involving the equivalent of hydrogen

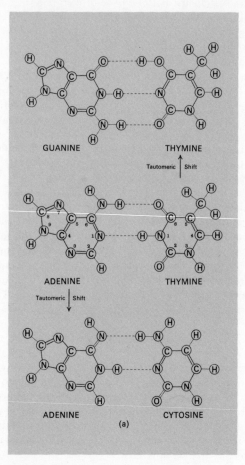

**Fig. 3.1**  The effect of tautomerism on base pairing.
(a) Adenine-Thymine is a standard base pair (centre). But a tautomeric shift in adenine involving the transfer of a hydrogen from C⁶ to N¹ would lead to A-C pairing (bottom).

Likewise, a transfer of hydrogen from N¹ of thymine to the C⁶ position would lead to G-T pairing (top). The two tautomeric forms would pair with each other.

atoms (protons and electrons). When these shifts involve those hydrogens at the sites of cross-bonding, other base-pair combinations become possible (Fig. 3.1).

Watson and Crick suggested that if an already incorporated base

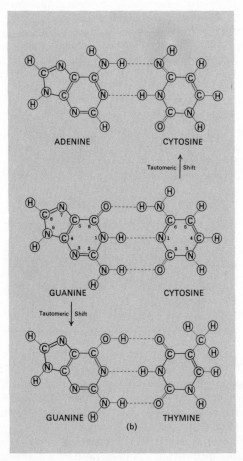

**Fig. 3.1** (contd)

**(b)** Guanine-Cytosine is a standard base pair (centre). But a tautomeric shift in guanine involving the transfer of a hydrogen from $N^1$ to $C^6$ would lead to G-T pairing (bottom).

Likewise, a transfer of hydrogen from $C^6$ of cytosine to the $N^1$ position would lead to A-C pairing (top). The two tautomeric forms would pair with each other.

existed in one of its less likely tautomeric states at the time of DNA replication, it would template for the 'wrong' partner. At the next replication cycle, when the two chains separated, the tautomer would probably revert to its more usual form and proceed to template in the normal way. But its 'wrong' partner, being a base in its most likely tautomeric state, would persist and, at replication, accept what is the 'correct' partner for it. The net result would be the substitution of one base pair for another. More specifically, a purine on one chain would be replaced by the alternative purine $(A \leftrightarrow G)$ while the pyrimidine of the partner chain would be replaced by the alternative pyrimidine $(T \leftrightarrow C)$; thus,

$$\text{Purine}_A \text{ --------- } \text{Pyrimidine}_T$$

$$\Big\Updownarrow$$

$$\text{Purine}_G \text{ --------- } \text{Pyrimidine}_C$$

Substitutions of this sort, in which the purine-pyrimidine orientations on the partner chains are maintained, are called transitions (Table 3.3).[34]

Table 3.3    Base-pair substitutions

| Transitions | G - C | C - G | A - T | T - A |
|---|---|---|---|---|
| ↑ | ↑ | ↑ | ↑ | ↑ |
| Original | A - T | T - A | G - C | C - G |
| ↓ | ↓ | ↓ | ↓ | ↓ |
| Transversions { | T - A | A - T | C - G | G - C |
| | | or | | |
| | C - G | G - C | T - A | A - T |

Notice also, that the direction of the change, $(A\text{-}T) \rightarrow (G\text{-}C)$ or $(G\text{-}C) \rightarrow (A\text{-}T)$, does not depend simply on which base is tautomerized. It depends also on whether the offending base is already in the DNA or in the precursor pool (see below). In the former event the reproductive mistake is described as an error of replication, in the latter, an error of incorporation (Fig. 3.2).

The ionization of a base at a critical pairing site also could have an effect comparable with that of a tautomeric shift and the deamination of certain bases could lead to mispairing as well (Fig. 3.3). Thus, amino groups are present in three of the four bases, namely adenine, cytosine and guanine. The first two are, in fact, the two 6-amino bases, and the 6 position is involved in both base-pair combinations. Guanine, on the other hand, is a 6-keto base but it does have an amino group at the 2 position. And, in the case of G-C pairing, a third hydrogen bridge is

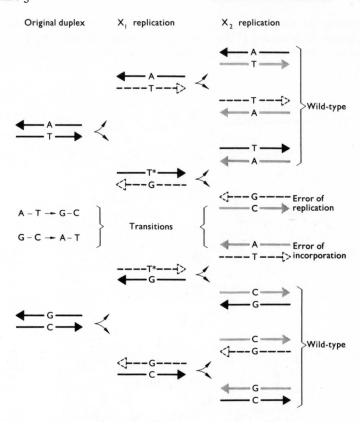

**Fig. 3.2**   Transitions.   (Top) At the first replication cycle (X$_1$), owing to tautomerization or ionization, thymine (T*) in the DNA pairs with the wrong purine from the pool (i.e. guanine). The sister helix is normal. At the second replication cycle (X$_2$) the thymine (T*) of the heteroduplex (G-T*) helix reverts to its normal (T) form and pairs with adenine. But the normal guanine in the sister molecule pairs with normal cytosine. The net effect is an A-T → G-C transition. (New strands shown below their respective templates.) (Bottom) At X$_1$, owing to tautomerization or ionization, a thymine (T*) base from the precursor pool pairs with a normal guanine in the DNA. At X$_2$ the newly incorporated thymine reverts to its normal form (T) and pairs with adenine. The net effect is a G-C → A-T transition. (New strands shown above their respective templates.)

formed between the 2 positions. Thus, deamination is expected to affect the pairing of all three amino-containing bases and, indirectly, that of the fourth, non-amino base, thymine. However, there are reasons for believing that the pairing properties of xanthine, the deaminated product of guanine, are similar to those of guanine rather than adenine. Thus, it is expected to pair with cytosine albeit by two hydrogen bridges only.

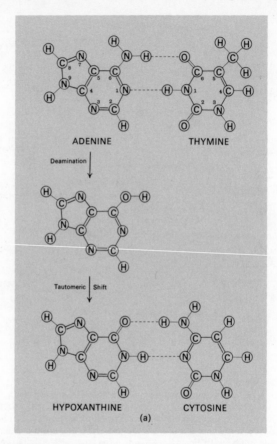

**Fig. 3.3**  The effect of deamination on base pairing.
**(a)** The final product of adenine deamination is complementary to cytosine.

These largely theoretical considerations provide a tentative basis to account for some spontaneous mutations and for the effects of certain mutagenic agents. For example, one of the ways in which nitrous acid, a known mutagen, is thought to act is by causing the oxidative deamination of adenine and cytosine. Thus, it has been found that pH affects the rate of nitrous acid-induced mutation in $T_4$ phage in the same way as it affects the rate of adenine and cytosine deamination. The pH effect on the killing caused by nitrous acid, on the other hand, follows the rate of guanine deamination (see p. 48).

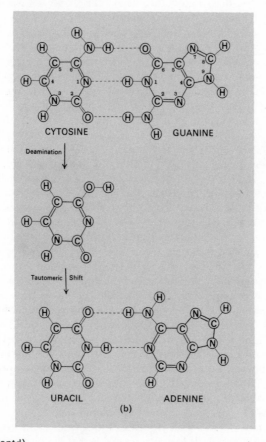

**Fig. 3.3** (contd)
**(b)** The final product of cytosine deamination is complementary to adenine.

In 1954 it was found that certain artificial analogues of the nucleic acid bases could be incorporated into the molecule, specifically replacing, to varying extents, the natural bases they resembled. Thus, as we have seen, thymine (5-methyl uracil) can be replaced by 5-bromo-uracil and 5-iodo-uracil. All of these have the same pairing properties because the 5 position is not involved in cross-bonding (but see p. 39).

For the moment, therefore, we need not regard the substitution of one base by an analogue as itself constituting a mutation. Indeed, again as we have seen, 'natural analogues' are a constant feature in many species. However, it has been found that when phages are grown under conditions which facilitate the incorporation of 5-BU, for example, the mutation rate increases with the degree of substitution. The reasons for this are not certain. Perhaps the analogue is more prone to tautomerization, or, more likely, ionization, than the normal base, a suggestion which can be extended to deaminated derivatives also.

In fact, little is known about the relative frequency of the tautomeric forms of the various natural or artificial DNA bases. But 5-BU does have a lower pK than thymine. Clearly, the absence of the $N^1$ hydrogen in the ionized (or tautomerized) form of either would allow pairing with guanine. Consequently, the lower pK of 5-BU may be important in relation to its enhancement of mutation rate. This factor may be significant also in regard to the direction of the transitions it is believed to induce. Thus, as indicated above, whether the change is (A-T) → (G-C) or (G-C) → (A-T) will vary according to the kind of error involved.

Now the pK of 5-BU is lower prior to polymerization than it is following incorporation into a polynucleotide. Consequently, the ionized form of 5-BU is expected to be more common in the precursor pool than in the template. On this basis (G-C) → (A-T) transitions, which arise by errors of incorporation, should exceed the opposite change which results from an error of replication (Fig. 3.2). On the model proposed for the mutagenic action of 5-BU, errors of replication should continue for as long as 5-BU persists in the template—even when free 5-BU is no longer available. These are expected to be (A-T) → (A-5BU) → (G-C) transitions. Errors of incorporation, on the other hand, leading to transition in the opposite direction, should require a continuing supply of 5-BU precursors. To this extent, therefore, the direction of transition can be recognized and, in keeping with the above-mentioned pK values, (G-C) → (A-T) transitions appear to predominate.

An important difference in regard to mutagenic action between base analogues and compounds like nitrous acid is that the former are actually incorporated into the DNA while the latter modify the structure of the bases. Consequently, the analogues are mutagenic only under conditions of DNA synthesis. But nitrous acid treatment of DNA prior to its replication can lead to mutation at a subsequent synthesis even though $HNO_2$ may no longer be present. For example, if the RNA fraction from tobacco mosaic virus is isolated, treated with nitrous acid and then allowed to infect tobacco leaves, a large proportion of the progeny virus contain altered protein. Likewise *in vitro* $HNO_2$ treatment of transforming principle or mature phage leads to mutation.

On the models proposed for the effects of mutagens such as nitrous acid and base analogues, the first step should be the production of so-called heteroduplex DNA in which one of the polynucleotide chains is normal at any one level while its complement is mutant or potentially so. This is easily shown for nitrous acid because *in vitro* treatment can be effected while no replication is in progress. But by supplying analogues towards the end of the phage replication period inside the bacterium, the transient heteroduplex state has been demonstrated in the case of analogue-induced mutations also. Of course, this involves the study of the progeny produced by single phage particles but this can be achieved (see p. 84).

Indirect support for the heteroduplex stage is seen also in the fact that this peculiar type of mosaicism does not occur in the case of $\phi$X174 the DNA of which is single-stranded.

If the basic mutagenic actions of hydroxylamine ($NH_2OH$), which is thought to change hydroxymethyl cytosine into uracil, and 5-BU, nitrous acid, etc. are the same and lead to $(A-T) \rightleftharpoons (G-C)$ transitions, then under appropriate circumstances, these mutagens should be able to reverse their own effects (back-mutation) as well as those of the others. This is frequently the case.

Similarly, 2-amino-purine, which has a low incorporation into DNA but which could replace adenine, induces mutations *in vivo* which are reversed at low rate by 5-BU and *vice versa*. But only about 10% of the spontaneous rII mutants of $T_4$ phage can be reverted by these mutagens, and few if any of the mutants induced by proflavine treatment are reversed by them.

It would appear, therefore, that transitions can account for only a minority of spontaneous mutations and different mutagens evidently effect changes of different kinds. The extent to which transversions occur is far from clear (see p. 175) but certain structural rearrangements, including loss or addition of one or more base pairs are known to occur. Indeed, it is very likely that mutagens like proflavine, an acridine dye, induce the deletion or insertion of bases (see p. 166).

Deletion may be involved also in some of the mutational pathways following treatment with alkylating agents. These compounds alkylate the $N^7$ of guanine and, less readily, the $N^1$ and $N^3$ of adenine and the $N^1$ of cytosine. Now alkylation of the guanine residue increases the rate of hydrolysis of its glycosidic linkage. The fracture of this link could lead to the rupture of the sugar-phosphate backbone. This break could persist, with subsequent loss, or errors could occur during the repair of the break or apurinic gap. However, ionization and tautomeric effects cannot be excluded and a further mutational pathway is possible which may be shared by both nitrous acid and UV irradiation. This involves the establishment of covalent linkages between bases at different but adjacent levels in opposite polynucleotide columns of a double helix, bases in the same column or else bases in different molecules. Some kind of inter-chain, intra-duplex cross linking is indicated when the strand separation normally observed on heating duplex DNA does not materialize. This effect has been found following UV treatment and thymine dimers (Fig. 3.4) are thought to be involved in this case. Difunctional alkylating

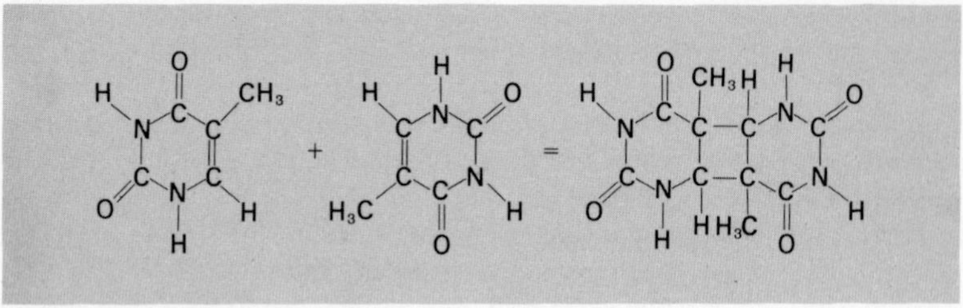

**Fig. 3.4**   The cyclobutane structure of a thymine dimer.

agents appear to be effective in this direction also for each reactive alkylating group can alkylate the $N^7$ position of a guanine. In this event two guanine residues could be linked by the C—C—N—C—C— sequence in the alkylating agent. This bridge of five atoms is, in fact, long enough to link guanine residues at adjacent levels in the same or different columns of the double helix.

Covalent cross links are expected to interfere with the replication of duplex DNA by impeding or preventing strand separation. Secondary consequences like breakage and deletion are expected to follow this interference. In fact, it has been shown that nitrous acid treatment can prevent complete separation of the strands and some of the rII mutants of $T_4$ phage induced by this chemical are undoubtedly large deletions.

Considerable uncertainty exists regarding the mode of action of ionizing radiations although the mutations induced by them have been studied for a long time. But as the above account illustrates even base analogues with their highly specific sites of incorporation cannot be held to have particular and unique effects. Consequently, much of the fore-going discussion regarding the mechanism of mutation induction must be regarded as provisional and even available evidence is not wholly consistent. Problems arise not only from a dearth of chemical information but from limitations in the biological methods which can be adopted for the detection and characterization of mutants.

# Interlude

## THE MAP CONCEPT

So far, we have considered some of the experiments which led to the recognition of DNA and, in some viruses, RNA, as the materials responsible for heredity. We have also seen something of the structure, mutation and replication of these molecules.

Obviously, the transformation experiments which showed that DNA was the genetic material in the bacteria employed, could not have been contemplated had it not been for the existence of genetically determined differences between related strains. In other words, differences owing to mutation must occur before heredity can be studied or even appreciated. What is more, mutations must affect metabolism, development, structure or behaviour (i.e. the phenotype), preferably in a conspicuous way, before their own nature and, hence, that of heredity, can be conveniently studied.

Clearly, this immediately raises a problem because phenotypic differences do not necessarily reflect genotypic differences, the environment also has its differentiating effects. In fact the phenotype is a product of interaction between the genotype and the environment. It is equally clear, therefore, that the occurrence of a mutation, or the presence of genetic differences owing to previous mutation or recombination, cannot be determined simply from phenotypic comparisons. Many criteria must be satisfied before phenotypic differences can be attributed to genetic change.[24] But the clearest evidence is provided by the fact that sites of mutation can be mapped relative to one another and the maps so produced correspond in some sense to the physical structures responsible for heredity. Various techniques can be used in this connection some of which are applicable only to special situations. But before these mapping procedures can be appreciated, the genetic systems of the organisms concerned must be understood.

The genetic systems of bacteria and those of their viruses are very intimately connected. Further, many of the properties of temperate viruses are shared by various other episomal elements (plasmids) including those which influence sexuality and gene transfer in bacteria. These connections and similarities have necessitated the inclusion of numerous cross references in the text and the reader, according to his existing knowledge, may find it more convenient to consider the various sections in a sequence other than that in which they are presented.

It is worth remembering also that the vegetative nuclei of higher plants and animals are produced by the mitotic division of the product of nuclear fusion (fertilization). They are, in consequence, diploid and the nature of the products of genetic segregation and recombination (meiosis) can usually be studied only through the diploid phase produced by their fusion.

In viruses, bacteria and most fungi, on the other hand, vegetative nuclei arise from the equational division of the products of a reduction division similar to, or analogous with, meiosis. They are, therefore, haploid and this means that the products of segregation and recombination can be studied prior to their fusion.

# PART II

---

## THE GENETIC ORGANIZATION
## OF THE GENOTYPE

# 4

# *The Transmission of the Genotype*

## THE GENETIC SYSTEMS OF BACTERIOPHAGES

DNA-containing bacteriophages are of two kinds with regard to the interaction which occurs between them and their hosts. On a number of occasions in the preceding sections, various aspects of bacteriophage life cycles have been considered.[18] All of these were related to what is called the lytic cycle which is invariably followed by virulent phages.

### Virulent phage and lysis

The lytic cycle consists essentially of:

1. The attachment of the phage to the cell wall of a sensitive bacterium;
2. The injection of the viral DNA into the host bacterium;
3. The take-over of the host synthetic machinery by the viral DNA;
4. The utilization of this machinery for the transcription, translation and replication of the viral genome;
5. The production of complete, infectious, progeny phages and their release following the death and lysis of the host bacterium.

This cycle is obligatory for virulent phages. But other bacterial viruses, described as temperate, may follow either the lytic cycle or an alternative one which establishes a state of lysogeny (Fig. 4.1).

3

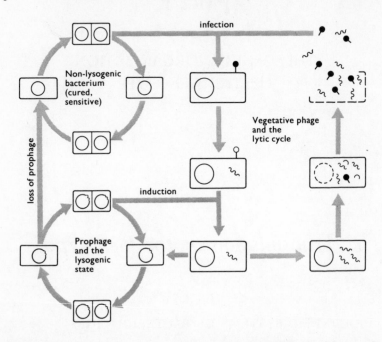

**Fig. 4.1** The lytic cycle and the lysogenic state. The former is obligatory for virulent phage but temperate phage are facultative in this respect. The bacterial genoneme is represented as a circle, that of the phage as a wavy line.

## Temperate phage and lysogeny

Soon after the bacterial viruses were discovered it was found that certain strains of bacteria were themselves a source of phage particles. Cultures of these bacteria always contained free phage particles which could not be eradicated even if:

1. Serial, single-cell sub-cultures of the bacterium were established;
2. Heat-resistant bacterial spores were heated to temperatures which were known to kill bacteriophages; or
3. The cultures were treated with phage anti-sera.

It now appears that this so-called lysogenic state is the rule rather than the exception. Generally a bacterium is lysogenized by only one kind of phage but double or triple lysogeny is not uncommon. The capacity of bacteria for releasing phages is associated with a second peculiar property, namely, resistance to lytic superinfection by the type or types of phage released. This immunity does not depend on an inability of the superinfecting phage to enter the lysogenic bacterium but on its failure to reproduce following entry.

Although phage particles which are infectious towards sensitive (non-lysogenic) bacteria are released by lysogenized bacteria, either spontaneously or following artificial induction (see below), infectious phages cannot be extracted from lysogenic bacteria. In them the virus exists in a latent state called prophage. Phages which can adopt this state are described as temperate as opposed to virulent.

The enigma regarding the source of the phage particles was partly resolved by a pedigree analysis of individual cells of lysogenized *Bacillus megatherium*. Single cells were cultured in micro-drops and as these divided their daughters were removed and cultured separately. This procedure was repeated for a large number of cell generations and, by this means, it was shown that:

1. Each bacterial cell gave rise to a lysogenic clone.
2. About 20 cell generations could elapse before any phage particles were released.
3. The release of phage followed the lysis of the bacterial cell.

Thus, it appears that the lysogenic state is a heritable property involving, in general, both the ability to release phage particles and resistance to lytic infection by the temperate phages responsible for lysogeny.[99] The first property, however, is expressed with only a low probability per bacterium per generation. This probability varies with the phage in question from about $10^{-2}$ to $10^{-5}$ but the value is fairly constant for a given phage under uniform conditions.

However, a variety of treatments (e.g. X-ray or UV irradiation, peroxide treatment) can induce the lytic cycle in many, but not all, lysogenic bacteria. Inducibility depends on the phage in question and not on the bacterium. Induced lysis follows after a lag of about one or two cell divisions and virtually all the bacteria in a treated culture respond to induction.

The first two stages described above in relation to the lytic cycle apply equally to the initial phases in the establishment of the lysogenic state. Similarly the later stages apply when the lysogenic state gives way to the lytic cycle and the prophage enters a vegetative phase. But, clearly, the establishment of lysogeny itself must involve a special and peculiar process. In fact crosses between sensitive and lysogenic strains using $F^-$ recipients and Hfr donors (see p. 65) have shown that the lysogenic versus the non-lysogenic difference is transmitted in essentially the same way as differences controlled by mutations of the bacterial chromosome. Further, crosses between bacteria which are lysogenic for different strains of the same phage have shown that certain prophages themselves segregate in the same manner as the bacterial genes. These observations can be explained on the same basis as that adopted in regard to the sex factor in Hfr strains. In other words, the establishment of lysogeny depends on the attachment or insertion of the phage particle into the bacterial chromosome. Experiments described later have shown that certain phages have very specific attachment sites (Fig. 4.2). These are capable of effecting only localized transduction (see p. 62). However, the temperate phages responsible for generalized transduction seem to attach at a variety of positions.

The replication of the integrated prophage, like that of the integrated sex factor, is synchronized with that of the bacterial genome. Further, the induction of the lytic cycle can be compared with the change from Hfr to F⁺. Thus, in its cytoplasmic state, the phage is released from host

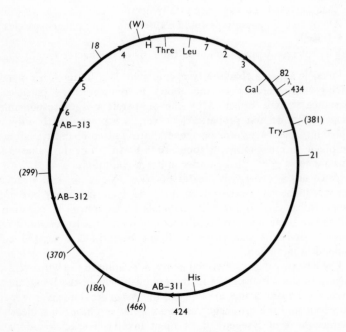

**Fig. 4.2** The circular linkage map of *E. coli* K-12. The arrows indicate the location and orientation of transfer of the integrated sex factor in the various Hfr strains which are symbolized inside the circle. The symbols outside the circle refer to the prophage locations of various inducible (in roman) and non-inducible (in italic) temperate phages. Some of these locations have not been exactly determined with respect to their neighbours (in brackets). Also shown are the locations of gene markers discussed later in the text (cf. Fig. 5.7).

control and replicates more rapidly. This cycle of reproduction is terminated by the lysis of the bacterium and the liberation of infective phage particles. On entry into a new sensitive host, these particles may enter a lytic cycle or a lysogenic cycle. However, a chromosomally-located prophage, transferred from a lysogenic Hfr donor to a sensitive F⁻ recipient, invariably causes lysis. This process is called zygotic induction (see p. 106). Thus, temperate phage particles transmitted by infection may adopt alternative states but prophages transferred by conjugation are obliged to enter a lytic cycle.

Phages have phenotypic effects on their lysogenic hosts which do not depend on their ability to transduce. Effects of this kind which do not result from transduction are called phage or lysogenic conversions.[7] A classic example of this conversion is found in the diphtheria bacillus, *Corynebacterium diphtheriae*. The harmful effects of this organism arise mainly from the production of a very poisonous exotoxin but this is not synthesized by all strains. However, phage preparations from toxigenic strains can confer toxigenicity on a proportion of the cells of non-toxigenic strains which survive the treatment. This conversion is correlated with the establishment of lysogeny and is associated with immunity to lytic super-infection. But the phage alone cannot be held responsible for toxin production. Thus, not only are certain strains of the virus concerned incapable of effecting the conversion, but non-toxigenic bacteria may harbour prophage which, on induction, can confer toxigenicity on other bacterial strains of the same species. Toxin production is therefore a venture in which both bacterium and virus participate. Clearly, the possibility of phage (lysogenic) conversion must be excluded before the prospect of transduction is entertained (see p. 60).

## THE GENETIC SYSTEMS OF BACTERIA[46]

### Transformation

As we have seen (p. 7), this phenomenon, first described by Griffith, was elucidated by Avery, MacLeod and McCarthy in 1944. It consists of the entry into recipient bacteria of naked DNA which is subsequently incorporated into the resident genome.[60] Recipients reject single-stranded DNA and the molecules in the transforming principle must be above a minimum size. It would appear also that recipient cells can take-up transforming DNA only during a particular phase of their development when they are said to be competent (see p. 92).

The nature of competence is not fully understood but the rejection of small molecules indicates that increased permeability is not involved.

One hypothesis suggests that competence depends on the development of enzymatically active receptor sites on the bacterial surface.[87]

Soon after transforming DNA is taken up, it begins to replicate at the same rate as the resident genome and it has a high probability of effecting transformation.[80]

## Transduction

In 1951, the Lederbergs and their co-workers found that characteristics of one strain of *Salmonella typhimurium* could be transferred to other strains at low frequency. The effect was observed even when the two strains in liquid culture were separated by a filter through which bacterial cells could not pass. Indeed, even cell-free extracts from the donor strain could effect the transfer and the efficiency of this principle was not destroyed by treatment with DNase. Thus, both conjugation (see p. 64) and transformation were excluded and the process was called transduction.

In fact, the transducing ability of the donor strain proved to be associated with its lysogenic state (see p. 56), the phage involved in the original case being PLT-22, more recently called P22. Thus, when P22 viruses grown on cultures of one strain infected other cultures of the same strain, no genetic conversions were observed. So-called phage or lysogenic conversion was thus excluded (see p. 59). But when distinguishable strains were used as recipients, such donor traits as motility, drug resistance, fermentation capacity and so on could be transferred. Markers in respect of which donor and recipient were similar were not affected. Thus, it is not the phage itself which causes 'directed mutation' of the recipient. A frequency of about one transductant per million phage particles is usually found in such experiments (LFT—low frequency of transduction lysates). Different characters are generally transduced independently but many cases of linkage are known (see p. 100). Further, they are usually transduced at about the same frequency but exceptions to this also are known and 20-fold differences in frequency have been recorded.

The plaque-producing capacity of a phage suspension can be described in terms of:

$$\text{Efficiency of Plating} = \frac{\text{Number of plaques produced (plaque titre)}}{\text{Total number of phage particles}}$$

Thus, the plaque titre is a function of the number of phage particles but the efficiency of plating and, hence, the plaque titre, varies also with cultural conditions. Thus, in the case of temperate phage (see p. 56) the lytic cycle and, consequently, plaque formation also, is favoured by conditions conducive to bacterial growth while lysogeny predominates when the temperature is lowered or protein synthesis is inhibited by chloramphenicol.

When the plaque titre of a transducing phage preparation is varied by concentrating the phage (by filtration or centrifugation), effectively diluting the phage by heat inactivation or changing the cultural conditions, the ratio between its plaque titre and its transducing titre remains the same. A phage is implicated more specifically by the fact that the ratio of plaque titre to transducing titre is constant also when both are depressed by the application of phage-specific anti-serum. Further, the number of transductants increases linearly with the multiplicity of phage, but a plateau or limit is reached which corresponds with the phage concentration which saturates the specific phage receptor sites on the bacterial cell, the number of which varies with the species. In fact, the attachment of P22 depends on the presence of host antigen 12, and *Salmonella* species which lack this antigen can neither absorb the phage nor are they transducible by it.

It is clear, therefore, that the transducing principle and the phage have similar specificities, but it does not follow that the phage particles which cause lysis or lysogeny and those which effect transduction are identical. In fact, the two capacities are usually distinct. Thus, not all species of *Salmonella* which are transducible by P22 (i.e. those with antigen 12) are lysed or lysogenized by transducing P22 phage suspensions. What is more, although lysogenic strains are immune to lytic super-infection (see p. 56) by the same or a related phage, they can nevertheless be transduced by them and, in this event, they retain the original prophage. These observations show that while the transducing element depends on a phage particle for its transmission, it is not identical with the normal infective phage.

In fact, a study of high frequency of transduction (HFT) preparations of λ-phage has shown that transducing phages (λdg) are different from normal phage particles. For example, initial transduction is achieved only at the rate of about one transductant for every million phage particles. But if the lysogenic colonies which are formed by the rare transductants produced by this initial transducing preparation are induced to enter the lytic cycle, the lysate yields a high frequency of transductants. Indeed, it would appear that about half the particles in this preparation are potential transductors. Now, if this HFT lysate is diluted to such an extent that a bacterial cell is very unlikely to absorb more than one particle, the resulting transductants are immune to super-infection and they can be induced to lyse. But the lysate does not contain transducing particles.

It appears, therefore, that while a transducing phage can effect transduction, it cannot give rise to infective particles. Consequently, the transducing capacity of HFT preparations must depend on the fact that they come originally from bacteria which were mixedly infected by both transducing and normal phage. Indeed in view of the low frequency of transducing phage in the original preparation the transductants produced can hardly escape simultaneous infection by normal phage. Thus, the deficiencies of the transducing phage are compensated for by the presence of normal phage in the same cell (see p. 179). Further, the high efficiency of the lysate derived from these mixedly infected recipients shows that the transducing particle persists and replicates in the original recipient and its descendants and the enhanced frequency of trans- duction reflects the high proportion of transducing phage in the prepara- tion. Indeed, since about half the phage in HFT lysates is able to transduce, it would appear that the initial transductant population produces approximately equal numbers of normal phage and so-called transducing phage.

The general view is that bacterial genes can be picked up by phage particles but these genes are incorporated at the expense of part of the original phage genome. This deletion of phage genes leads to defects which vary in kind and extent depending on the nature and the length of the lost segment. For example, in the case of P1 phage, some trans- ducing particles appear to be normal and capable of producing progeny which can both transduce and cause lysis without a subsidy from normal phage. But, at the other extreme, some transducing particles of the same phage are so defective that they do not even confer immunity to super- infection. There are reasons for believing that the bacterial segment picked up by a phage is, at least approximately, of constant length. But the phage segment which is absent from the transducing phage may be longer or shorter than this. Thus, each isolate, in a sample of ten in- dependently derived transducing preparations of λ-phage, was found to have a characteristic density. Some were up to 8% heavier than normal lambda while others were as much as 14% lighter.

The transductions discussed above are those of the complete variety because, although a small proportion of the transductants may lose their transduced character (cf. prophage and sex factor and see p. 65), the vast majority gives rise to stable transduced clones. Two patterns of complete transduction are known depending on the nature of the phage: P22 can transduce small segments from any part of the bacterial genome (unrestricted or generalized transduction), while λ-phage can transduce only a particular short segment of the host chromosome (restricted or localized transduction). These two patterns differ in other respects also. For example, in the latter, transducing phage is produced only after the induction of lysogenic bacteria; the lysate produced spontaneously fol- lowing lytic infection is ineffective. But in the former both preparations are capable of transduction.

Transductions are not always of the stable type, however, and in the case of the generalized transducer P22, for example, many abortive transductions are achieved for every complete transduction effected (Table 4.1). In abortive transduction, the transducing particle enters the recipient cell and the donor bacterial genes it carries function in a normal way. Consequently, the recipient cell shows the donor character. But the injected DNA is unable to replicate, so that, when the recipient cell

**Table 4.1**  Frequencies of abortive transduction for various markers in *Salmonella typhimurium*. (Data of Ozeki, H. (1959). Chromosome fragments participating in transduction in *Salmonella typhimurium*, *Genetics*, **44**, 457–70, after Hayes, W. (1965) *The Genetics of Bacteria and their Viruses*, Blackwell, Oxford.)

| Recipient strain (bracketed strains contain linked markers) | No. transductants per $4 \times 10^8$ phage P22 particles | | Ratio abortive : complete |
|---|---|---|---|
| | abortive | complete | |
| ade C-7 | 1120 | 81 | 13·7 |
| adth C-5 | 1240 | 105 | 11·8 |
| gua A-1 | 1030 | 190 | 5·4 |
| try D-10 | 3810 | 346 | 11·0 |
| cys B-12 | 3430 | 503 | 6·8 |
| try D-10, cys B-12 | 3070 | 136 | 22·5 |
| ser-1 | 1440 | 160 | 9·0 |
| ser-5 | 1260 | 152 | 8·3 |
| his D-39 | 18880 | 1988 | 9·5 |
| adth A-4 | 7160 | 644 | 10·8 |
| pro A-46 | 1760 | 376 | 4·7 |
| cys A-1 | 1700 | 188 | 9·0 |
| met C-50 | 900 | 64 | 14·0 |
| adth D-12 | 480 | 52 | 19·2 |

divides, the transducing DNA passes to only one daughter. Thus, the recipient cell gives rise to a colony only one cell of which is transduced. In fact, virulent phage may exist temporarily in a comparable state of abortive lysogeny.

The evidence suggests that complete transduction depends on the attachment or insertion of the transducing DNA into the bacterial chromosome (integration). In this state the transducing phage genome, like that of a normal prophage or integrated sex factor, reproduces in concert with the bacterial chromosome. But if integration is not achieved, the transducing DNA, unlike a normal vegetative phage or sex factor, is unable to replicate, transduction is abortive and transmission of the transduced feature is unilinear.

Phage such as lambda which effect localized transduction have a particular site of attachment and they are able to transduce only those bacterial genes which are closely linked to their preferred location (see p. 58). Phage which can bring about generalized transduction, on the other hand, have variable and transient attachment and, consequently, they can carry a variety of bacterial genes.

Temperate phage and the sex factor (see p. 65) have many common properties although the latter cannot be transmitted autonomously by infection. It comes as no surprise, however, that following its integration in Hfr strains, the F factor can pick up bacterial genes. Recipients which receive these substituted sex factors (F') following conjugation may be converted to the donor phenotype. In acknowledgement of the similarities involved this process has been called sexduction. It accounts for the occasional occurrence of recipients which receive certain donor genes much earlier in the conjugation process than expected on the basis of linear transmission of the chromosome from Hfr strains. Thus F' donors have properties which are intermediate between those of normal F$^+$ donors and Hfr strains. Consequently, they are known as intermediate donors and F' shows the same orientation of transfer as the Hfr strain from whence it came. The location of the sex factors varies from one Hfr strain to another. Consequently, although sexduction is localized for any one strain, it is generalized when various strains are considered.

## Conjugation

The existence of conjugation and recombination in bacteria was discovered by Lederberg and Tatum in 1946. Subsequently it was found that recombination was mediated by a one-way transfer of genetic material during the mating process, one cell (F$^+$) behaving as a donor, the other (F$^-$) as a recipient.[42] The polarized nature of the transfer is expressed in the consequences of mild UV irradiation. Thus, donor-treatment enhances recombination as much as 50-fold but recombination is much reduced by irradiation of the recipient. This polarity has been confirmed in various other ways, for example by the examination of individual ex-conjugants following their separation by micro-manipulation and from a study of labelled DNA transfer. Despite the donor/recipient relationship, both ex-conjugants survive after mating because most bacteria under standard conditions of culture are 'multi-nucleate'.

The F$^+$ and F$^-$ states are relatively stable, they breed true and, though donors sometimes mutate to the recipient state, the reverse change does not occur. This and other evidence indicates that the change from a donor to a recipient state is due to mutation by loss rather than substitution.

The so-called sex factor (F) which determines the donor state has several unusual properties:

1. When $F^+$ and $F^-$ types are incubated together for about an hour in broth, between 5% and 95% of the $F^-$ recipients are converted into $F^+$ donors. This is in sharp contrast to the behaviour of more conventional genetic elements.
2. When an $F^+/F^-$ mixture contains a large excess of recipients, it can be seen that they are converted into donors at a faster rate than the bacteria divide.
3. Cell-free extracts from donor bacteria cannot convert recipient strains to the $F^+$ state.
4. Treatment of $F^+$ cultures with low concentrations of acridine orange leads to the loss of the sex factor. The donors are thus converted into recipients but remain otherwise unchanged.

These observations are consistent with the view that the sex factor of $F^+$ strains is a 'cytoplasmic' genetic entity whose replication, like that of a virulent virus, need not be synchronized with that of the host cell. But unlike a virus it is not autonomously infective. It is, however, transferred during conjugation thus effecting a recipient → donor conversion.[49]

Subsequently, however, it was found that $F^+$ donor strains could change not only to $F^-$ but to donor forms which produce a high frequency of recombinants for certain markers. These so-called high frequency of recombination (Hfr) strains differ from $F^+$ donors in many respects. For example:

1. In $F^- \times F^+$ matings the frequency of recombinants is low and of the same order of magnitude for all genes.
2. In $F^- \times Hfr$ matings however, some gene markers are transferred with a frequency no greater than that found in $F^- \times F^+$ matings. But certain other markers, the identity of which varies from one Hfr strain to another, are transferred a thousand times more efficiently.
3. Recipients which receive markers transferred at low frequency by Hfr strains are often converted into Hfr donors but those which receive only the markers transferred with high frequency almost invariably remain recipient.
4. The breeding behaviour of recipients which have been converted into Hfr donors is the same as that of the original Hfr donor.
5. Hfr strains arise only from $F^+$ lines, and Hfr strains readily revert to the $F^+$ state.

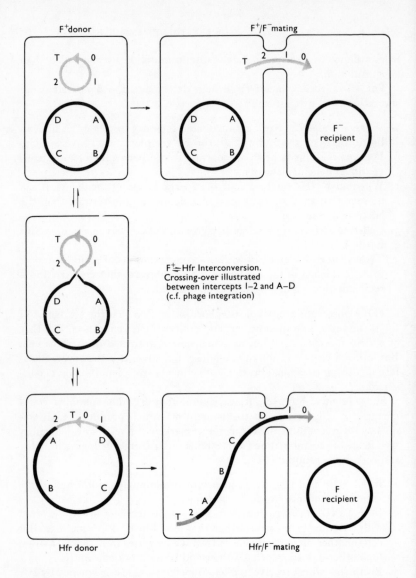

**Fig. 4.3** The interconversion of F⁺ and Hfr donor states in bacteria. The evidence indicates that both the sex factor and the bacterial genonemes are circular in the vegetative state but are transferred linearly during conjugation. The sex factor determines the orientation of transfer in both F⁺ and Hfr donors. Note that, unless recombination occurs between O and T during integration, the sex factor is transferred in two sections (leading and trailing) by Hfr donors. Presumably a complete sex factor is required for full donor function. Hence this is the last character to be transferred by Hfr donors.

The fact that F⁺ and Hfr represent alternative, mutually interchange-able states introduces a difficulty in relation to genetic analysis. Thus, when an $F^+ \times F^-$ mating is set up, one cannot be certain that all the donors are in the F⁺ state at the time of conjugation because some of them may have adopted an Hfr state in the meantime. In fact, it would appear that much, if not all, the transfer of genetic material, other than that of the sex factor itself, from F⁺ to F⁻, depends on the presence of a low frequency of Hfr forms in the donor culture.

Further, the evidence indicates that during its transfer from Hfr donors the bacterial chromosome behaves as a linear structure although it appears to be circular in the vegetative cell (see p. 27). What is more, the fact that only certain markers are transferred with high frequency during this process indicates that:

1. The point of breakage in the circle is constant for a given Hfr strain.
2. A particular end of the linear structure leads the way during transfer (see Figs 4.2 and 4.3).
3. Generally, only a part of the genome is transferred.
4. The segment transferred is of variable length.

Jacob and Wollman offered an ingenious hypothesis to account for these particular phenomena. They suggested that the sex factor could exist in alternative states. In F⁺ donors it was held to occur as an extra-chromosomal or cytoplasmic particle. In this state it itself was efficiently transferred during conjugation thus effecting a recipient-to-donor con-version. But its cytoplasmic status in F⁺ donors did not allow it to mediate in the transfer of chromosomal bacterial genes except in special circumstances (see p. 64). In this state it was capable of replicating in-dependently of the donor chromosome but was susceptible to attack by such agents as acridine orange.[26]

On this basis, the F⁺ → Hfr transition was supposed to involve the insertion or attachment of the sex factor at some point on the bacterial chromosome. In this position it replicated in synchrony with the chromosome and was transmitted by the same process of oriented partial transfer as that responsible for the passage of normal bacterial genes.[2] But since the sex factor is rarely transmitted during chromosome transfer, it, or part of it, seems to occupy the anchor position. Whether the position of the break determines the point of attachment of the sex factor or whether the point of attachment determines the locus of breakage is not clear (but see Fig. 4.3). What is certain, however, is that the break-point and, hence, the orientation of the transfer, does vary from one in-dependently isolated Hfr strain to another (see p. 107). The similarity between the behaviour of the sex factor in F⁺ and Hfr donors and that of temperate phages in their alternative lysogenic and lytic cycles is very similar (see p. 58). In fact, the amount of DNA in the sex factor, as estimated from its rate of inactivation by the decay of incorporated $^{32}P$, is approximately $2\cdot5 \times 10^5$ base pairs which is about the same as that in DNA of bacterial viruses.[73]

A particle of this size is expected to carry enough information to determine a number of properties. In fact, it appears that the various aspects of donor function are largely, if not entirely, the responsibility of the sex factor. Interestingly, both $F^+$ and Hfr donors are attacked by RNA bacterial viruses while $F^-$ recipients are resistant. Their immunity depends on the absence of specific receptor sites. Donors and recipients differ also in their surface antigens (cf. phage or lysogenic conversion, see p. 59).

## THE GENETIC SYSTEMS OF FUNGI

In most organisms, cell fusion occurs only in relation to sexual reproduction and the fusion of gametes is soon followed by a fusion of gametic nuclei. These nuclei may fuse before or during the first division to follow gametic fusion. In either event the binucleate stage initially created by gametic fusion is confined to the zygotic cell. Many fungi conform to this general pattern but the group also provides many exceptions to this standard scheme (Fig. 4.4).

### Dikaryosis

In filamentous members of the ascomycetes and basidiomycetes, the fusion of gametic cells is not followed closely by a fusion of their nuclei. Rather the two nuclei of the zygotic cell divide mitotically and synchronously on separate spindles so that four nuclei, two from each source, are produced. These mitoses and subsequent wall formation are organized in such a way that the two resulting cells each receive a mitotic product from both gametic nuclei. Thus, the two binucleate cells produced are identical with each other and with the zygotic cell. This sequence of synchronized mitoses may continue until a more or less extensive tissue of genetically identical binucleate cells is produced. This tissue is called a dikaryon. In the ascomycetes, the dikaryotic phase is developed only within the fruit body (perithecium) which forms around the female organs (ascogonia). But in basidiomycetes the dikaryon is more extensive as it often shows considerable vegetative growth. Indeed, the dikaryon is the principal vegetative phase of many basidiomycetes and can be cultured on a synthetic medium.

In both groups, the two nuclei in cells of the dikaryotic phase eventually fuse to give a truly diploid nucleus which divides meiotically to produce, in most cases, uninucleate asco- or basidio-spores as the case

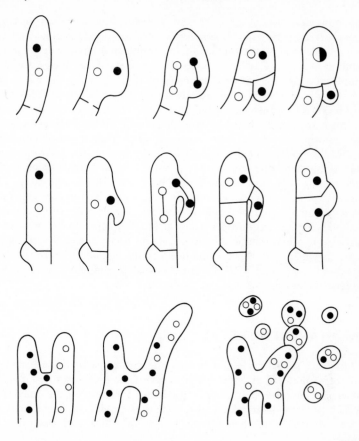

**Fig. 4.4** Nuclear cycles in fungi. (Top) The dikaryon and the production of a diploid nucleus via crozier formation in ascomycetes. (Centre) The perpetuation of the dikaryotic phase via clamp connections in basidiomycetes. (Bottom) The formation of a heterokaryon and its perpetuation and partial breakdown via conidia.

may be. Many fungi, including members of the ascomycetes and basidiomycetes have self-incompatibility mechanisms so that fusion leading to a dikaryon can occur only between gametic cells which differ in respect of all the genes involved in the determination of mating type.

## Heterokaryosis

The cells which serve gametic functions in those ascomycetes which have been used extensively in genetical research are generally specialized and uninucleate (monokaryotic). But the vegetative mycelium in types such as *Neurospora* and *Aspergillus* is only irregularly septate and the segments defined by the septa contain variable and often widely varying numbers of nuclei. Further, the septa are perforated and nuclei as well as cytoplasmic material can pass from segment to segment. Of course, the mycelium produced by a uninucleate sexual or asexual spore will contain only one kind of nucleus (homokaryon). But fusion can occur between the vegetative parts of two or more genetically-different homokaryons. The resulting mycelium, containing various proportions of two or more genetically-different nuclei, is called a heterokaryon. In *Neurospora*, at least, stable heterokaryons can form only between homokaryons of the same mating type, and certain other genes must be common also. The various types of nuclei which contribute to a heterokaryon are generally recovered unchanged through uninucleate or homokaryotic conidia. However, nuclear fusion does occur at low frequency in heterokaryons and what is called mitotic crossing-over can occur in the diploid nuclei so produced. Indeed, this may be followed by haploidization so the final outcome is similar to that of normal sexual reproduction. In fact, the sequence is known as the parasexual cycle.

## THE STANDARD SEXUAL CYCLE

The basic sexual cycle of cellular organisms consists of the fusion in pairs of haploid (n) gametic nuclei to give diploid (2n) zygotes. Zygote nuclei, or some of their mitotic products, then undergo a reduction division (meiosis) so that haploid nuclei are re-produced. These, or their mitotic descendants, then serve gametic functions and the cycle is complete.

This basic cycle is interrupted and extended by the occurrence of mitosis in the haploid phase, in the diploid phase, or in both haploid and diploid phases (Fig. 4.5). These mitoses may be concerned with the development of the two phases or with their asexual reproduction.

The genetically important features of meiosis are, in sequence:

1. The specific pairing of apparently single-stranded, genetically-equivalent or homologous chromosomes derived ultimately from opposite gametes.
2. The splitting of these paired chromosomes to give two pairs of sister chromatids. These genetically-identical sister chromatids remain in parallel and associated in the region of the centromere. This is a special structure which is responsible also for the movements of the chromosomes on the division spindles.

3. The accurate usually reciprocal exchange of corresponding segments between non-sister chromatids in the paired associations (bivalents). One or more exchanges of this type can occur per bivalent. Any pair of non-sister chromatids can be involved in exchange but only two

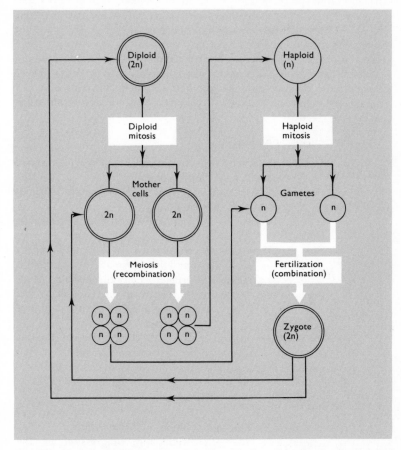

**Fig. 4.5** The relationship between mitosis, meiosis and fertilization in standard sexual cycles.

chromatids participate in any one recombinational event. Exchange of this kind is called crossing-over and its occurrence is associated with chiasma formation. As the term indicates, a chiasma is a cruciate structure which forms because, owing to the exchange, the pairs of chromatids on one side of it and those on the other are differently constituted with regard to longitudinal continuity though not origin (Fig. 4.6).

4. The co-ordinated but randomly oriented separation of the double chromosomes in each bivalent. Orientation and separation occur in relation to a spindle-shaped structure of oriented protein fibrils, and they result in the production of two daughter nuclei. Each of these contains only half as many chromosomes as the original zygote but each chromosome is two stranded.

5. The co-ordinated but randomly oriented division of the chromosomes in each daughter nucleus gives, in all, four meiotic products. These contain the haploid number of single-stranded chromosomes. For ease of reference this sequence is generally considered in terms of the following stages:

    I.  First meiotic division:

       (a) Prophase: Leptotene—the appearance of the single-stranded chromosomes.

                 Zygotene—the pairing of homologous chromosomes.

                 Pachytene—the splitting of chromosomes and the occurrence of crossing-over.

                 Diplotene—the stage when the double-strandedness and the chiasmata become visible.

                 Diakinesis—the stage when chromosome arms rotate relative to one another so that the chiasma assumes its definitive 'cross' shape.

       (b) Metaphase I—the orientation and congression of the bivalents on the first division spindle.

       (c) Anaphase I—the separation of homologous centromeres.

       (d) (Telophase I—the formation of two daughter nuclei).

    II.  Second meiotic division:

       (a) (Prophase II—The condensation of chromosomes).

       (b) Metaphase II—the orientation and congression of the chromosomes on the equator of the second division spindles.

       (c) Anaphase II—the splitting and movement apart of half-centromeres.

       (d) Telophase II—the formation of daughter nuclei.

Telophase I and prophase II have been put in brackets because they may be omitted. Thus, in their mechanically-active state the chromo-

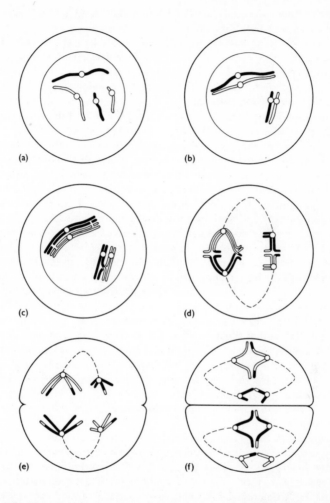

**Fig. 4.6** Meiosis. (a) Leptotene: Two pairs of homologous chromosomes are shown which differ in length and arm ratio (i.e. centromere position). Maternal and paternal contributions are distinguished by colour. DNA synthesis has been completed by this stage but the replication products have not been individualized as distinct chromatids. (b) Zygotene: Specific pairing. (c) Pachytene: Chromatid formation followed by breakage and reunion of non-sisters to give crossing-over and chiasmata. (d) Metaphase I: Random co-orientation and congression of bivalents. (e) Anaphase I: Separation of homologous centromeres and their associated chromatids. (f) Met-Anaphase II: Division of centromeres following auto-orientation and congression.

somes are compact and spiralized while they tend to be diffuse and extended in metabolically-active, non-dividing nuclei. The condensation required by division is effected during prophase and reversed during telophase. But the contracted chromosomes of first anaphase may proceed directly to the second division as soon as new spindles are organized. In this event both telophase I and prophase II are side-stepped. But if resting nuclei are reconstituted at the end of the first division, a period of condensation is required in preparation for the second division. In other words, if telophase I is included, prophase II is required.

The halving of the chromosome number is an inevitable consequence of this double-division sequence because, although the nucleus divides twice, the chromosomes divide only once. By the same token, although a particular segment or locus is represented twice in the zygote and its mitotic products, it can be represented only by a single dose in the meiotic products and their mitotic descendants. This process is called reduction, or, since qualitative differences may be involved, segregation.

Loci which are subject to alternative representation in the haploid phase are said to be genetically allelic and the mutually-exclusive forms are known as alleles. Loci at strictly corresponding positions on homologous chromosomes must, of course, be allelic because they cannot both occupy the same strand. But even if two loci do not correspond exactly they will still behave as alleles in transmission if, for some reason, crossing-over cannot separate them. For example, crossing-over does not occur at meiosis in male *Drosophila* and so whole maternal and paternal chromosomes are presented as alternatives for the gametes. Further, alleles are expected to be equally represented in the gametes of a heterozygote. Thus where $a$ and $\alpha$ represent an allelic difference, meiosis in an $a\alpha$ diploid cell should produce equal numbers of $a$ and $\alpha$ cells (but see p. 124).

Meiosis may also re-assort such non-allelic differences as may have distinguished the original gametes and for which the zygote was consequently heterozygous. This process is called recombination and it is achieved in two ways. First, the random orientation of maternal and paternal centromeres at first metaphase and the random orientation of chromosomes at second division serve to assort gene differences on non-homologous chromosomes. Second, crossing-over between paired homologues results in the recombination of non-allelic genetic differences on homologous chromosomes.

We have described in detail the relationships between crossing-over, orientation and recombination in an earlier publication (p. 226). In this book we will consider only recombination between homologous chromosomes, for this permits mutant sites, and thus genes within a linkage group, to be mapped in relation to each other (p. 110).

# EXTRACHROMOSOMAL HEREDITY

The genetic systems of bacteria and those of their viruses are so intimately related that the distinction between heredity and infection becomes somewhat blurred. This means, of course, that the distinction between the genetic material of the 'host' and that of the 'parasite' is by no means clear-cut. For example, the sex factor (see p. 65) is generally regarded as an integral, if not always integrated, part of the bacterial genome largely because its transmission depends on bacterial division and conjugation. But, as we have seen, the sex factor and temperate phage share many common properties, particularly in their integrated states in Hfr and lysogenic strains. Further, the episomal status of such particles denies the distinction between chromosomal and extra-chromosomal genetic elements.[79]

One of the early arguments in favour of the nuclear location of genetic elements was the fact that male and female gametes generally made equal contributions to the genotypes of their offspring even when their cyto-plasmic contributions were grossly unequal. But once the nature of chromosomal heredity had been elucidated it became clear that many patterns of transmission did not conform with those expected on the basis of known chromosome behaviour.

For example, some traits showed strictly or preponderantly maternal inheritance so that consistent and persistent differences obtained between the results of reciprocal crosses. Non-Mendelian and even somatic segregations were also observed and it was found that asexually produced offspring did not always breed true. These and other symptoms led to the conclusion that while the chromosomes were undoubtedly the principal vehicles of heredity they were not the exclusive sites of self-replicating genetic elements.

The accuracy of chromosome replication and movement provides the precision of Mendelian heredity. However, certain species or strains have anomalous, though characteristic, genetic systems and consequently their breeding behaviour departs from that expected on classical Mendelian grounds. For example, in males of the dung fly *Sciara coprophila* the paternally-derived chromosomes are eliminated at meiosis and they transmit, *en bloc*, only the genes derived from the maternal parent. This is true also of the males of certain homopteran insects but here the paternal chromosomes are also inactivated during development. Thus, on the basis of their appearance and breeding behaviour, the males appear to be the products of haploid parthenogenesis. But cytological study shows that this is not the case. In *Rosa canina*, on the other hand, the egg cell transmits four times as many chromosomes as the male gamete and a comparable situation is known in cecidomyids and earthworms.

In these and other ways the chromosomes can determine patterns of inheritance which are not usually associated with them. Further, many non-genetic influences can distort development so as to give spurious indications of extrachromosomal inheritance. But the following sample of case histories, taken from a variety of organisms, are typical of those which are clearly or most simply explained on the basis of extrachromosomal genetic elements of one kind or another.

## Sex ratio in *Drosophila*

Some natural populations of many species in the *obscura* group of the genus *Drosophila* include what are called sex-ratio males. On mating these produce 90% or more daughters whatever the genotype of the maternal parent. This sex-ratio condition is determined by the nuclear genotype. In fact, it behaves as though the controlling element was located in the right arm of the X chromosome though multigenic control is probable.

It would appear that only two of the meiotic products are functional on the male side even in normal *Drosophila* and the above condition seems to depend on the preferential movement of the female-determining X chromosome to the functional products.

A sex ratio condition of a quite different kind has been described in various other species of this genus including *D. willistoni*, *bifasciata*, *prosaltans*, and *equinoxialis*. Here too progenies consisting mainly or entirely of daughters are produced by certain strains, but the condition (SR) is transmitted only through the female. Thus, only rarely does the cross SR female × Normal male produce viable males, but when these are mated with normal females, the resulting sons and daughters are normal in character and relative frequency.

The low frequency of sons from SR females depends on the early death of male embryos. The agent responsible for this embryonic death can be extracted not only from the inviable embryos but from the somatic tissues of SR females. When this agent is introduced into normal females, they proceed to breed in the manner of SR females. In fact in *D. willistoni* the agent has been identified as a spirochaete and it does not appear to be species specific.

## Kappa in *Paramoecium*

So-called killer strains of *Paramoecium aurelia* secrete a substance known as paramecin to which they are resistant but which kills non-killer, sensitive strains. Killer and sensitive strains can, however, be mated and various treatments (e.g. heat, X-rays and chloromycin) can convert killers into non-killer sensitives (cured-killers).

Now, conjugation in *Paramoecium* is such that the two exconjugants have identical micronuclear genotypes and, under normal circumstances, there is no significant mixing of cytoplasms during the process.

Following the conjugation of killer and cured-killer cells from true breeding lines, one exconjugant of each pair produces a killer clone while its partner gives rise to sensitive cells; each exconjugant follows its parental cytoplasm. If, however, the cytoplasmic bridge which forms during conjugation persists for some time after nuclear exchange has been completed, both exconjugants may give rise to killer clones. The frequency with which killer fails to segregate increases with the duration of contact.

Clearly, the difference between killers and cured-killers is purely extrachromosomal. It depends, in fact, on the presence in killer cells of DNA-containing cytoplasmic kappa particles about $0 \cdot 4 \, \mu$ long whose existence had been postulated before they were seen. Cured-killers can be reconverted into killers also by acquiring kappa particles from a concentrated, cell-free extract from killer individuals.

Thus, kappa is usually transmitted only by cell division but it can be transferred naturally via conjugation if cytoplasmic contact is prolonged. Further, at least under laboratory conditions, it can be transmitted by artificial infection.

Killer strains vary in respect of the kind and amounts of paramecin they produce which suggests that kappa is subject to mutation. In fact, a given killer cell may carry more than one type of kappa particle. Indeed, various other cytoplasmic particles are known in *Paramoecium* which share many common properties and, perhaps, true homology with kappa.

The two situations outlined above can be regarded as abnormal in that the observed extrachromosomal determinants do not characterize the species concerned. Further, there is no reason to suppose that these determinants share any homology with the genetic elements, nuclear or otherwise, of the standard genome. In fact, both conditions can be attributed with some confidence to foreign organisms or elements. Thus kappa could be compared with a *Rickettsia* virus, which it resembles in general appearance, though its capacity for autonomous infective transmission is much lower than expected of a typical virus. In this respect, however, it resembles the sex factor of bacteria which, as we have seen, shares many common properties, except cell-free infectivity, with temperate bacteriophages. What is more, deciding whether the sex factor or temperate phage shares homology with the genome of the 'host' is more a matter of semantics than of science. Be these as they may, on the basis of their hereditary transmission as revealed in breeding experiments, there is little to distinguish SR and kappa from each other or from what we regard as normal cytoplasmic organelles with genetic continuity.

## Plastids in plants

The genetic continuity of extrachromosomal elements was first postulated by Correns on the basis of the non-Mendelian heredity shown by green-white variegated plants of *Mirabilis jalapa*. He found that the seeds from flowers on green parts of the plant gave rise to green seedlings, those on white parts generally gave white seedlings while those from variegated shoots gave green, white and variegated offspring with widely varying relative frequencies. These relations obtained whatever the source of the pollen. In other words, inheritance was strongly matrilinear.

Variegation in plants has many causes, but, for situations like that in *Mirabilis*, Baur suggested that green-white mosaicism depended on the occurrence of two types of chloroplast, normal green and mutant white, both of which showed genetic continuity. Thus, from a cell containing a mixture of such plastids (mixed celled), daughter cells could be produced which contained only one or other plastid type. The frequency of this segregation would depend on the number, relative frequency and degree of mixing of normal and mutant plastids in the mixed cells. Plastid segregation followed by cell segregation would give pure green and pure white sectors and shoots.

However, mixed cells have not been observed in *Mirabilis* though they have been seen in other cases of green-white variegation e.g. *Primula sinensis*, *Nepeta cataria* and *Spirogyra triformis*. In fact, Correns believed that mixed cells were not common enough for Baur's hypothesis to be generally valid. Consequently, he argued that the effect on the plastids did not depend on their own inherent nature (plastogenes) but rather on that of other, unseen genetic elements in the cytoplasm (plasmagenes). But however one interprets these situations in detail, the non-Mendelian nature of transmission is clear.

Chloroplasts are now known to contain DNA, and although its exact function is not clear, it is an obvious candidate for a genetic role.[82] However, even if plasmagenes can be shown to exist it does not follow that all plastid effects are due to plastogene mutation. In fact, this is known not to be the case for, as with mitochondria, many nuclear mutations are known to affect chloroplast structure and functions (see p. 212).

Strong, even exclusive, matrilinear transmission of plastid differences is a common phenomenon but in *Nepeta cataria* about 30% of the offspring receive plastids via the pollen while in *Pelargonium zonale* 70% of them do so. Lower rates of male transmission have been described in *Oenothera* and *Epilobium*.

## Petite in yeast

The so-called 'petite' mutants of yeast grow more slowly and produce colonies which are smaller than those of wild-type strains. However, this difference is observed only in the presence of oxygen, the petites having no measurable oxygen consumption. These observations suggest that the aerobic respiratory mechanism of the petites is in some way defective. This conclusion is supported by the following facts.

1. Petite cells are insensitive to cyanide and certain other inhibitors of aerobic pathways.
2. They lack several respiratory enzymes such as cytochromes $a$ and $b$, and cytochrome oxidase (Fig. 4.7). Cytochrome $c$ reductase is absent also while cytochrome $c$ is present in excess.

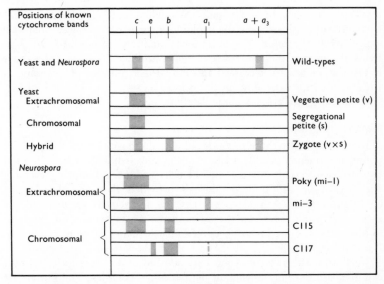

Fig. 4.7 Absorption bands of cytochromes from mitochondrial mutants in yeast and *Neurospora*. Band width is an indication of concentration. (After Wagner, R. P. and Mitchell, H. K. (1964), *Genetics and Metabolism*, John Wiley and Sons.)

3. The mitochondria, in which the above enzymes together with those of the citric acid cycle are normally contained, are abnormal in petite mutants in that they seem to stay in an incompletely developed state.
4. Petite mitochondria do not stain with Janus green.

Although the various petite mutants are phenotypically similar, they are of different kinds genetically. For example, the normal Mendelian pattern of segregation which obtains following the crossing of certain petite mutants (segregational petites) with wild-type shows that the defect depends on a single, chromosomal gene mutation. But the abnormal genetic behaviour of other petite mutants shows that their defects must depend on a change in some extrachromosomal element with genetic continuity. The petites which give indications of extrachromosomal heredity are themselves of two kinds, neutral and suppressive.

### Neutral petites

These are sexually sterile amongst themselves but they can be mated with wild-type haploid cells of opposite mating type. The zygotes thus produced can divide mitotically to produce diploid colonies by budding. Meiosis in these diploid cells produces four ascospores and an examination of these shows that the mating-type gene and other nuclear gene markers segregate in a typically Mendelian manner. However, the frequency of petites among the haploid segregants is no higher than that expected on the basis of spontaneous mutation from wild-type. This holds also even if the non-petite haploids produced following the above hybridization are back-crossed to petite for a number of generations. Clearly, therefore, the abnormality is in some way cured by the crossing, and nuclear genes are not immediately involved.

### Suppressive petites

In some respects these mutants show the opposite effect. Suppressive petites are phenotypically similar to the neutrals but their breeding behaviour is different. All the diploid cells produced following crossing with wild-type will undergo meiosis provided they are placed on a sporulation medium as soon as they are formed. A majority of the ascospores produced under these circumstances is petite. But the ratio of normal:petite produced by a given meiosis may vary from 4:0 to 0:4. Nuclear markers, on the other hand, again show normal segregation.

Further, if the original zygotes are allowed to divide mitotically, they can give rise to either normal or petite diploid colonies. The former are fertile and produce wild-type haploids by meiosis. But the latter are sterile and cannot now produce ascospores. What is more the ratio of normals to petites among the diploid zygotic products varies widely and continuously from nearly nought to almost a hundred per cent. This variation is affected by the conditions of culture and the strains employed.

In all the above cases there are reasonable and, sometimes, compelling grounds for associating the extrachromosomal genetic elements, inferred from the results of breeding experiments, with visible cytoplasmic particles. In fact, in the case of centrioles and other (homologous?) fibre-producing organelles the evidence for genetic continuity rests almost entirely on their physical continuity.

However, breeding experiments have provided good indications for the extrachromosomal transmission of many differences for which there is no obvious material basis. This is true for certain types of male sterility in *Zea* and *Epilobium*, for streptomycin resistance in the isogamous *Chlamydomonas reinhardi*, and for a variety of effects in fungi which have been extensively studied in this direction.

Many non-genetic factors, such as maternal nutrition and antigen incompatibility, carry-over gene effects and delayed gene activity, can give effects which mimic those of extrachromosomal inheritance. Consequently, especially where visible particles are not in evidence, very rigorous criteria must needs be satisfied before extrachromosomal heredity can be concluded.

## $CO_2$ sensitivity in *Drosophila*

Certain strains of *Drosophila melanogaster* depart from wild-type in being abnormally sensitive to high concentrations of carbon dioxide. Crosses between sensitives and wild-type show reciprocal differences, as follows:

sensitive ♀ × normal ♂ → true breeding sensitive males and females.

normal ♀ × sensitive ♂ → true breeding sensitive, and true breeding normal males and females.

In the second cross, the ratio of normal to sensitive is variable, non-Mendelian and independent of sex. Clearly, an extrachromosomal basis is indicated for the sensitive state. This is supported by the fact that normal flies are sensitized following the receipt of an implant or extract derived from sensitive donors. Further, in the case of female (but not male) recipients, sensitivity is transmitted to a fraction of the offspring. These, whether male or female, subsequently behave and breed in the manner of spontaneously sensitive strains. The condition has been attributed to an unseen, extrachromosomal element called sigma and temperature treatment, presumably by destroying or inactivating sigma, can convert sensitives to resistants.

Certain sensitive strains are peculiar in that sensitivity is shown only by the older individuals. The young flies are $CO_2$ resistant and immune to 'infection' by extracts from sensitive flies, and an infective principle cannot be extracted from them. Such a principle can be extracted, however, as soon as the flies become sensitive. The condition in young flies is reminiscent of the lysogenic state of bacteria which carry temperate viruses in the prophage form while the onset of sensitivity and associated phenomena can be compared with the switch from the prophage to the vegetative state. The element responsible has been regarded as a homologous mutant form of sigma called rho. Various other homologous mutants have been defined on the basis of deviant properties. For example, some $CO_2$ sensitive strains show very low rates of male transmission and the element responsible has been called omega.

## The red variant of *Aspergillus*

*A. nidulans* can be propagated in various ways; sexually via uninucleate, haploid ascospores and asexually by either uninucleate, haploid conidiospores or hyphal tips containing many haploid nuclei produced by mitosis. The colonies produced from single spores are expected to contain nuclei of only one kind (homokaryotic) and asexually-produced sub-cultures of such colonies are not expected to vary.

However, sectoring during development, and segregation during asexual reproduction, have been observed for a variety of characteristics in this and many other fungi. For example, the red variant of *A. nidulans* is developmentally stable but while some of its asexually-produced conidiospores give rise to red colonies, others germinate to give a normal (non-red) mycelium. The mitotically-produced spores of the red segregants invariably repeat this segregation pattern and they give between 10% and 90% of normal segregates. This somatic segregation has been shown to persist for over ten years during which time tens of thousands of sub-cultures have been produced involving over three hundred cycles of asexual, single spore propagations. The spores from segregants of normal type also can give rise to normal and red colonies but the frequency of the latter does not exceed 10% and true breeding normals can be isolated from the red segregating strain. Essentially similar results are obtained when hyphal tips are used for sub-culturing.

Heterokaryons can be established between the red-variant and non-segregating normal strains which differ in respect of various nuclear gene markers. From these heterokaryons, lines can be isolated which carry the nuclear genes of the normal parent but which show the red phenotype and segregating character of the red variant.

The behaviour of the red variant can be explained on the following assumptions:

1. It is a persistent heteroplasmon containing a mixture of mutant and normal homologous plasmagenes (cf. mixed cells in green-white variegation).
2. Individuals containing only the mutant form of the plasmagene are inviable (cf. albino plastids).

On this basis the frequency of segregation and the ratio of segregants produced depend on the absolute and the relative numbers of the two homologous plasmagenes.

Ascospore samples which give rise to a high frequency of red colonies are less viable than those which yield lower frequencies of red colonies. This suggests that cells with high relative frequencies of mutant plasma-genes are less vital than those with lower frequencies which is in keeping with the second assumption given above.

If this is so then, other things being equal, the mutant plasmagene should gradually be selected out and the red variant should revert to the non-segregating normal type. Of course, the experimenter is constantly selecting in the opposite direction, purposely perpetuating the variant. But there is another aspect to this situation.

Conidiospores from the red variant may give rise to colonies which are initially only slightly different from wild-type; presumably these contain a preponderance of the normal plasmagene. But as they grow these colonies become fully mutant in appearance and successive samples of conidiospores taken from them give rise to progressively increasing frequencies of red colonies. These observations suggest that, when it is outnumbered in the heteroplasmon, the 'red plasmagene' reproduces at a faster rate than its normal homologue. In other words, the fitness of the mutant plasmagene is inversely related to its frequency. This is a common basis for the maintenance of polymorphism at the level of both genes and of individuals.

# 5

## Mapping the Genetic Material

### MAPPING IN BACTERIOPHAGE

A cross in the case of vegetative phage consists in simultaneously infecting sensitive bacteria with two (or more) genetically-marked strains of the same phage. Unabsorbed phages can be removed by centrifugation or treatment with phage-specific anti-serum. The progeny phage produced following lysis are then examined by plating on suitable bacterial hosts. If small samples are taken from highly diluted suspensions of the originally-infected bacteria before they lyse, the phage produced from single host cells can be studied separately (single burst experiment).

An examination of the progeny phage from a population of bacterial cells shows that doubly different parental phages give rise to reciprocal recombinants in equal numbers. The joint percentage frequency of these recombinants represents a recombination value which can be used in the construction of a linkage map. However, the conditions of the experiment must be strictly standardized for otherwise reproducible results are not obtained. By this method recombination values as low as 0·01% have been detected and problems associated with merozygosity do not arise.

It will be appreciated, however, that experiments of this kind do not study the outcome of a single generation of crossing—even when single bursts are studied. Rather a population produced by about five or six replication cycles is examined and both parental and recombined types may be involved in more than one cycle of recombination. The relationship between observed recombination values and idealized mapping functions can be reasonably discussed for cellular organisms because the general properties of meiosis and fertilization are understood. But a comparable discussion in relation to phage requires a knowledge of the mating and recombinational processes together with information on the pattern of genesis of mature infective phage.

On the basis of certain assumptions regarding these events, Visconti and Delbruck formulated a mating theory in which the kinetics of phage crosses are expressed algebraically.[89] This expression allows a definition

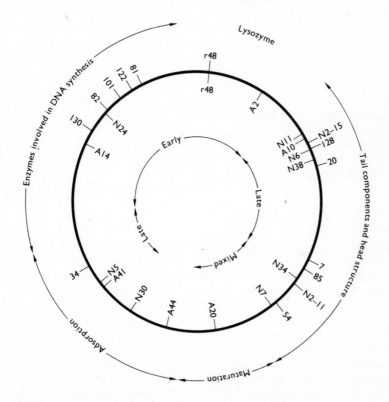

**Fig. 5.1**  The circular linkage map of T$_4$ phage. The symbols inside the map refer to temperature sensitive mutants which form plaques at 24°C but not at 42°C. Those on the outside refer to amber mutants which form plaques on a certain (amber suppressor) mutant strain but not on wild-type strains of *E. coli*. In all, about 80 genes have been distinguished and mapped but only about 20 proteins are produced by the phage. The broad distribution of gene function is indicated outside the map and the broad distribution of early-acting versus late-acting genes is shown inside it. Note clustering of genes of similar function. The genoneme itself is a linear structure but the linkage map is circular because individual particles represent circular permutations of the structure represented. Ten degrees on the map are equal to c. 20 map units.

of the relationship between the observed recombination value, on the one hand, and, on the other, the average number of matings, the incidence of recombination and the frequency of the various parental-type genes in the mating pool. Further, additive linear mapping functions can be derived from the formulae which compensate for the occurrence of multiple exchanges and multiple rounds of mating.

The experimental data agree well with the predictions of this mating theory but it will not be discussed in detail because some of the assumptions on which it is based are not likely to be true, and different assumptions can give equally acceptable predictions.

However, two features of phage recombination are of particular interest. First, series of two- and three-point crosses similar to those discussed later have shown that the linkage maps of the T-even phages are circular (Fig. 5.1). Second, although reciprocal recombinants are equally represented among the phage produced from a population of bacteria, there is strong evidence to suggest that reciprocal products do not arise from a single recombinational event. It would appear therefore, that there are fundamental differences between the mechanisms of recombination in phage and in higher organisms.[85]

## MAPPING IN BACTERIA

Genetical studies in bacteria are peculiar because the zygote itself can be studied only rarely. In fact, even its production can often be ascertained only posthumously and the events which occur during and following its formation must, in general, be determined from an examination of the recombinants derived from it. Of course, in relation to higher organisms with diplontic cycles it could be said that the outcome of meiosis can be determined only by an examination of the diploid phase, while in haplontics the meiotic events which occur within the zygote are revealed in its haploid products. But in higher organisms one generation is easily distinguished from the next and the haploid and diploid phases can hardly be confused. Further, with special exceptions, the two parents make an equal contribution to their offspring and parental-type progeny are easily distinguished from the parents themselves. Furthermore, meiotic recombination in cellular organisms generally gives balanced reciprocal products and, except in special circumstances, all recombinants are expected to survive.

However, a common feature of all the semi-sexual cycles in bacteria is that the transformed, transduced, sexduced or exconjugated recipient is only partially diploid (merozygous). This complicates mapping in many ways. First single or any odd number of exchanges between donor-fragment and recipient result in two incomplete and probably inviable reciprocal recombinants except in the special case where the donor fragment corresponds with a terminal region of the recipient genome (if such exists) or where donor or recipient genonemes are circular. Consequently, strong natural selection against the products of odd numbers of exchanges is expected. Second, of the two reciprocal products which result from a given, even-numbered series of exchanges between linear entities, only one has a complete genome. Therefore, both products of a given series are not likely to be viable.

Added complications arise from the fact that bacteria are multinucleate and so even the zygotes are chimaeral.[11] What is more a donor fragment may be involved in more than one cycle of recombination. For these reasons a single recipient cell may give rise to more than one type of daughter and in no way can it be decided whether those cells having the character of the recipient strain came from 'zygotic nuclei', non-zygotic nuclei within zygotes, or whether they are simply recipient cells which did not receive a donor fragment at all.

Apart from the complications which are inherent in the genetic system, there are other operational difficulties. These arise largely from the selective techniques the investigator must employ to detect and isolate recombinants especially when these arise at very low frequency.

Hayes has summed-up the situation by saying that 'in bacterial systems, we are quite unable to assess, even on a statistical basis, the number of zygotes involved in a cross or the total number of parental and recombinant progeny which issue from them, so that the results we obtain cannot be expressed in terms of classical recombination frequencies, much less map units'. But what results can one obtain?

The methods employed are various but, in all cases, the frequency of recombinants obtained in respect of one pair of markers is compared with that observed in respect of another pair. In essence, these values are obtained as follows. A suitable cross is first established and, where appropriate, the donor cells are destroyed after zygotes have been formed. The cells of the recipient/zygote population are then allowed to grow. Those products of this population which are purely recipient in character are ignored. This is a necessity because, as indicated above, there is no way of telling which of them are recipient-type products of zygotes and which are unmated recipients. Certain recombinant classes may be ignored also, not because they are irrelevant but because a convenient selective technique for isolating them cannot be applied. In fact, positive selection is practised in respect of those recombinants which have received a particular gene from the donor. And this fraction alone is then tested to see how many of them have also inherited other donor genes. These other genes are described as unselected markers because the recombinant population actually studied was isolated without reference to them.

Some justification for the method can be seen in the following formal model. Let us suppose that recipients *abc* receive the donor fragment *ABC*. Let us further suppose that the 'real' recombination values for the marked intercepts are 10% and that this is also the average recombination value for both the segment to the left of *A* and that to the right of *C*. Recombinants selected for the *A* donor marker are expected to be of four types. Their mode of origin and their expected relative frequencies are shown in Table 5.1. If these expected frequencies are expressed as a percentage of the total number of *A*-carrying recombinants, we get $ABC = 33·2\%$, $ABc = 33·2\%$, $Abc = 33·2\%$ and $AbC = 0·4\%$. Thus, among the recombinants selected for the *A* marker we expect to find that *B* is present in 66·4%, *C* is present in 33·6% and, of those with *B*, 50% have *C* also. It can then be argued that the 'recombination values', i.e. the percentage frequency with which linkages are broken, are:

$$A \times B = 33·6\% \qquad A \times C = 66·4\% \qquad B \times C = 50\%$$

Thus, the *AC* distance appears to be twice the *AB* intercept and the following map, in round fractions, could be produced

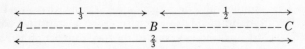

Notice that, in contrast to the situation in conventional systems, the value for the *BC* distance obtained by subtraction $(\frac{2}{3} - \frac{1}{3} = \frac{1}{3})$ is more realistic than that obtained directly, the two differing, not by twice the 'expected' frequency of double recombinants, but by the expected frequency of double recombinants itself $(\frac{1}{2} - \frac{1}{3} = \frac{1}{6} = \frac{1}{2} \times \frac{1}{3})$. This is because exchange at the 'two-stranded' stage has been considered.

However, this nice correspondence between the three recombination values is somewhat spurious and depends on the fact that the joint occurrence of *B* and *C* was determined by selecting for *B* first and selecting for *C* subsequently. In other words the order in which the donor genes were selected corresponded with their order on the chromosome. But the investigator has no way of knowing the sequence until the experiment has been performed! Consequently, one is equally justified in selecting for *B* among the recombinants which contain *C*. Reference to Table 5.1 shows that if this course is adopted, the *CB* types are expected to constitute 99·8% of the *C*-carrying recombinant population. On this basis, we might be tempted to claim that *B* and *C* are very closely linked and that this linkage is broken in only 0·2% of the cases. This

**Table 5.1** The production of *A*-carrying recombinants following *ABC* → *abc* transfer. (+) exchange present; (−) exchange absent.

Merozygote and recombination values

```
        A   B   C
     ...  ...  ...
      10  10  10
     ---  --- ---
      10  10  10
      ...  ...
       a   b   c
      (−)     (−)
```

| Viable *A*-containing recombinants | Minimum exchanges required | | Probability | |
| --- | --- | --- | --- | --- |
| | Pattern | No. | Absolute | As % of *A*-carrying |
| *ABC* | A B C<br>+ − − +<br>a b c | 2 | $0.1 \times 0.9 \times 0.9 \times 0.1$<br>$= 0.81\%$ | 33.2 |
| *ABc* | A B C<br>+ − + −<br>a b c | 2 | $0.1 \times 0.9 \times 0.1 \times 0.9$<br>$= 0.81\%$ | 33.2 |
| *Abc* | A B C<br>+ + − −<br>a b c | 2 | $0.1 \times 0.1 \times 0.9 \times 0.9$<br>$= 0.81\%$ | 33.2 |
| *AbC* | A B C<br>+ + + +<br>a b c | 4 | $0.1 \times 0.1 \times 0.1 \times 0.1$<br>$= 0.01\%$ | 0.4 |

would be quite wrong, of course, and the recombination values, $A \times B = 33 \cdot 6\%$, $A \times C = 66 \cdot 4\%$ and $B \times C = 0 \cdot 2\%$, are not additive and so do not yield a linear map. However, in this case too, equal values for the $AB$ and $BC$ intercepts can be obtained by subtraction.

Further, in the present example, where the four types of $A$-carrying recombinant are presumed to be distinguishable, there is no real problem because the very low frequency of $AbC$ recombinants shows that these are the 'double recombinants' so that the order must be $ABC$. Indeed, the conflicting recombination values obtained for $B \times C$ according to the selection sequence are themselves compatible. This can be shown by constructing matrices to fit the observed values.

Thus, when the frequency of $B$-carrying recombinants which also carry $C$ is determined, the three values obtained are, approximately, $B:b = 2:1$, $C:c = 1:2$ and $BC:Bc = 1:1$. These ratios are accommodated by the following unique matrix which shows that the $AbC$ class is virtually non-existent and must, therefore, represent the 'double recombinant' fraction. On this basis the order must be $ABC$. If, on the other hand, the frequency of $C$-carrying recombinants which also carry $B$ is determined, the three values obtained are, approximately,

```
          B │ b
          2 │ 1
        ┌───┬───┐
BC  1   │ 1 │ 0 │ 1   C
        ├───┼───┤
Bc  1   │ 1 │ 1 │ 2   c
        └───┴───┘
```

$B:b = 2:1$ and $C:c = 1:2$ as before, and, in contrast, $CB:Cb = 1:0$. These ratios too, however, are accommodated by the same unique matrix, as shown, so the two sets of results, though apparently in conflict, are actually mutually consistent. In fact, the converse relationships would be obtained if one could select for $c$ among the $b$ recombinants and for $b$ among the $c$ recombinants.

```
          C │ c
          1 │ 2
        ┌───┬───┐
CB  1   │ 1 │ 1 │ 2   B
        ├───┼───┤
Cb  0   │ 0 │ 1 │ 1   b
        └───┴───┘
```

Where two of the mutations involved in the cross affect the phenotype in the same way (e.g. confer susceptibility to the same drug or dependence on the same metabolite) certain recombinants cannot be distinguished one from the other. Under these circumstances the sequence can still be determined but the results of reciprocal crosses have to be compared. The rationale of this technique is considered later in regard to mapping by transformation (see p. 97).

To some extent, therefore, the maps obtained are comparable with those produced for organisms with a standard meiotic system. But while operationally useful maps can be constructed in the above manner, the method presupposes that the absence of a donor marker among the re-combinants depends entirely on the exchange pattern in the zygote (post-zygotic exclusion). Yet, clearly, in view of the fragmentary nature of the initial transfer, a particular, unselected marker ($B$ or $C$ in the above example) could be absent from a recombinant because the recipient did not receive it in the first place. In fact, results similar to those above would be obtained if the segments received were incorporated *in toto* but, owing to random breakage in their production (pre-zygotic exclusion), not all segments carrying $A$ carried $B$ or $C$ as well. For example, if the order is $ABC$ and the $A$ to $B$ and the $B$ to $C$ intercepts are equal, breaks in the donor are expected to separate $A$ from $B$ as often as they separate $B$ from $C$ while $A$ and $C$ should be separated about twice as often. Under these circumstances about two thirds of the $A$-carrying recipients and their descendants are expected to contain $B$ while only one third are expected to carry $C$. On this basis, however, $AbC$ types would not be possible unless two fragments, $A$-carrying and $C$-carrying, were separately incorporated. The fact that such types are produced under conditions of linkage indicates that exchange in the zygote does contribute to the recombination pattern. What is more, so far as map production is concerned it does not make much difference whether the assortment frequencies depend on breakage before transfer or recombination after receipt. In fact, both processes are probably similar. The question of pre- and post-zygotic exclusion will be reconsidered later in relation to bacterial conjugation. But enough has been said for the present regarding the peculiar nature of mapping methods in bacteria.

## Mapping by transformation

As we have seen, the first real indication that DNA is the genetic material in bacteria came from a demonstration of its transforming ability.[84] Thus, very pure and highly polymerized DNA extracted from one strain can effect permanent hereditary changes in another. The proportion of potential recipients which is transformed for any particular feature tends to be rather low and varies with the concentration of DNA in the transforming principle, at least over low concentrations. It varies also with the organism and its physiological state. Thus, in *Pneumococcus* as many as 10% of the recipients may be converted for a specific marker but values under 1% are more usual for *Haemophilus*.[71]

The molecular weight of the DNA in transforming principles is about $5-15 \times 10^6$. This is equivalent to about 0·2–0·5% of the total donor genome (c. 2–5 $\mu$). This means that a given DNA fragment in the transforming principle can cover only a comparatively small segment of the recipient. Consequently, a given segment is expected to transform two characters simultaneously only if the mutations which determine the character-differences concerned occupy adjacent or closely linked sites. This is the rationale for using joint transformation as a basis for mapping mutant sites but, owing to the relatively small size of the donor fragments, only short regions of the genome can be considered at any one time (but see p. 99).

At first sight it might appear that linkage could be claimed if the frequency of doubly transformed cells exceeded the product of the frequencies of cells transformed for each character considered separately. Thus, it could be argued that if exposing *ab* recipients to transforming DNA from *AB* donors gives 5% transformants for *A* and 5% transformants for *B*, the frequency of double *AB* transformation, on the basis of independent events, is expected to be 0·25%. On this basis values significantly above 0·25% could be regarded as an indication of linkage between *A* and *B*. This kind of argument is valid in many cases, genetical and otherwise. But there are reasons why it cannot be applied directly in cases of transformation.[39] These reasons cannot be discussed in detail but the inappropriateness of the above reasoning is seen in the relationship between the frequency of transformants obtained and the concentration of DNA in the transforming principle. Thus, at low DNA concentrations, these show a simple linear relationship which demonstrates that transformation can be effected by a single molecule of transforming DNA. A simple relationship also obtains between the frequency of transformation and the amount of DNA absorbed by recipients. However, increasing the DNA concentration above about 10 molecules per bacterium does not result in any further increase of transformants. What is more the limiting frequency is comparatively low, never exceeding 10% and usually much lower. It appears that this limit reflects a restriction on the number of DNA molecules which a recipient can absorb, and bacteria are in a competent state only during a limited period of the cell and culture cycle. Thus, an individual bacterium is receptive to transformation only for a period equal to about half that of the cell generation time and the yield of transformants in a culture is highest when it is exposed to transforming DNA towards the end of the period of exponential growth.

The spurious linkage which may appear when only a fraction of the recipient population is receptive can be illustrated as follows. Let us suppose that *A* and *B* transformants, when considered separately, each constitute 5% of the recipient population. If all the recipient cells are competent, then double *AB* transformants are expected with a frequency of 0·25% on the basis of independent events. But if only half the cells of the recipient culture are competent at the time of exposure to transforming DNA, the *A* and *B* transformants each represent 10% of the competent cells. Consequently, the bacteria which are doubly transformed for unlinked genes are actually expected to constitute 1% of the competent fraction. And since the competent fraction represents half the total recipient population, double *AB* transformants are expected to constitute 0·5% of the total recipient population—even when $A$ and $B$ are unlinked. This is twice the value calculated on the assumption of 100% competence. But, clearly, an observed value of 0·5% cannot be taken as evidence of linkage. Further, the disparity increases as the fraction of competent cells decreases.

The complications and errors introduced by variations in competence and in the efficiencies of transforming principles can be avoided. The rationale behind the method which establishes linkage relationships unambiguously becomes clear when transformation is considered in terms of the fragments of DNA which bring it about.

Thus, the question to ask in relation to linkage is this—are the double $AB$ transformants produced following the integration of two separate particles, one carrying $A$, the other carrying $B$, or are they produced by a single transforming particle which carries both $A$ and $B$?

This question can be answered. We have already pointed out that

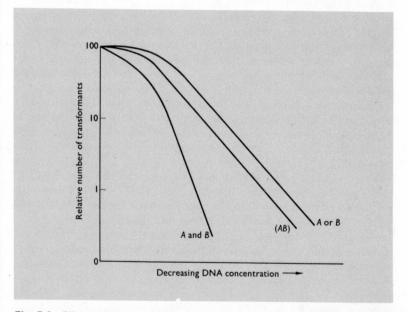

**Fig. 5.2** Effect of DNA concentration on frequency of transformants (idealized curves). The curve on the right shows the effect in relation to any single marker gene (e.g. $A$ or $B$). The centre curve shows the effect with regard to double transformants when the two markers ($A$ and $B$) are closely linked (i.e. can be carried by a single fragment). The curve on the left shows the effect on double transformants when the two markers are not linked and hence carried by different fragments. (Based on Goodgal, S. H., 1961, *J. gen. Physiol.*, **45**, 205–27.)

there is a linear relationship between the yield of single transformants and the concentration of DNA in the transforming principle so long as this is below saturation level. Now, if double transformation, like single transformation, is effected by a single particle then, clearly, the *rate* at which the success of double transformation falls with decreasing DNA concentration should be the same as the *rate* of decrease for single transformation.

If, on the other hand, double transformation depends on the co-incidence of two different transforming particles in the same cell, its success will decrease much more rapidly. Thus, if on dilution the probability of *A* transformation is decreased by a half, and the probability for *B* is likewise reduced by a half, the joint expectation of *AB* transformation is reduced by a quarter, and so on (Fig. 5.2).

Evidence that linked transformation really does depend on the physical association of the genes involved comes from studies on the loss of transforming activity following heating. Heating to what is called the melting-out temperature leads to the separation of the two chains in duplex DNA. This separation is accompanied by a loss of transforming ability. By carefully controlling the temperature rise it is possible to inactivate one gene before another. But while unlinked genes may be inactivated at different temperatures, linked genes tend to be inactivated at the same temperature (Fig. 5.3). For example, the ability of a preparation to confer microccin resistance on a sensitive strain is destroyed by heating to 88°C, but a temperature of 91·5°C is required in the case of amethopterin resistance. These genes are not linked. On the other hand two closely linked genes affecting streptomycin and sulphonamide resistance both lose their transforming ability at 89·5°C. And these two genes are not linked to the other resistance genes mentioned above. Now, the melting-out temperature is a function of the G-C content. These bases are joined in the double helix by three hydrogen bonds while only two are formed between A and T (see p. 17). And the stability of the duplex increases with the number of H-bonds. Thus, the fact that linked genes, unlike unlinked genes, tend to be inactivated at the same temperature suggests that they are carried by the same particle. If the inactivation really does depend on G-C content, these results also show that the frequency distribution of the base pairs is asymmetrical as between different segments of DNA each of which contains as many as 50 genes or so.

Linkage can be detected in transformation experiments by the method outlined above. Estimating the strength of linkage, on the other hand, introduces further complications for the general reasons touched on earlier. But it can be done. Thus, in *B. subtilis* mutations affecting the ability to synthesize histidine (and thus to grow in the absence of a

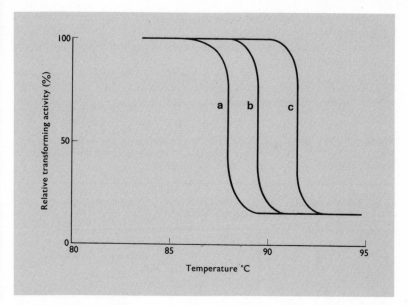

**Fig. 5.3** The effect of temperature on the transforming ability of DNA in respect of three markers in *Pneumococcus*. The ability to transform resistance to (a) micro-coccin is lost at 88°C, (b) streptomycin and sulphonamide, determined by distinct but closely linked genes, is lost at 89·5°C, and (c) amethopterin is lost at 91·5°C. (Data of Roger, M. Redrawn from Sager, R. and Ryan, F. J., (1961), *Cell Heredity*, John Wiley and Sons.)

histidine supply) are closely linked with those related to tryptophan synthesis. A 2-factor transformation experiment can, therefore, be performed with a view to mapping these loci in relation to each other. This is done by studying reciprocal transformation between *try⁺ his⁻* forms and *try⁻ his⁺* types. For example, in the case of a *try⁻ his⁺* recipient (Table 5.2), the expected detectable transformants are *try⁺ his⁺*, *try⁺ his⁻*, and *try⁻ his⁻*.

The production of *try⁺* descendants must depend on the co-incidence of:

1. recombination to the left of the *try* marker, and
2. recombination to the right of the *try* marker. This may be to the right or the left of the *his* locus.

All the *try⁺* products can be selectively isolated by growing the recipient/ transformant population on a medium containing histidine but no tryptophan. These *try⁺* transformants are clearly of two types (see 2 above) namely:

1. those in which the exchange to the right of *try* occurred in the *try-his* intercept (i.e. *try⁺ his⁺*), and
2. those in which the exchange to the right of *try* was to the right of *his* also (i.e. *try⁺ his⁻*).

These two types can be selectively isolated by transferring the *try⁺* transformants (isolated as indicated above) to a medium lacking histidine.

**Table 5.2** The exchanges involved in the formation of the detectable transformants obtained when *try⁻ his⁺* recipients are transformed by DNA from a *try⁺ his⁻* donor.

| Transformation | Detectable transformants | Exchanges required |
|---|---|---|
| | *try⁺ his⁺* | *try⁺ ──── his⁻* (exchange X) (exchange X) over *try⁻ ──── his⁺* |
| Donor: *try⁺ ──── his⁻*  Recipient: *try⁻ ──── his⁺* | *try⁺ his⁻* | *try⁺ ──── his⁻* (exchange X) ···· (exchange X) over *try⁻ ──── his⁺* |
| | *try⁻ his⁻* | *try⁺ ──── his⁻* (exchange X) (exchange X) over *try⁻ ──── his⁺* |

The frequency of exchange between the *try* and *his* markers can then be expressed as a function of that to the left of *try*. This, in turn, is a function of the length of the *try-his* intercept but, clearly, the 'map unit' involved is not comparable with that in conventional systems. A series of two factor crosses of this type, involving different pairs of linked markers, can provide information on the sequence of the markers involved. Thus, the data in Table 5.3 are consistent with the order *ant-try-his*. However,

**Table 5.3**  The results obtained in a series of 2-factor transformation experiments involving a total of three loci affecting different and distinguishable aspects of tryptophan and histidine synthesis in *Bacillus subtilis*. (Based on data of Anagnostopoulos, C. and Crawford, I. P. (1961) *Proc. natn. Acad. Sci. U.S.A.*, **47**, 378–90.)

| Cross | Recipient | | Donor | | Frequency of $+\,+$ transformants as a percentage of $+\,\pm$ transformants | 'Mean' |
|---|---|---|---|---|---|---|
| 1a | *try*⁻ | *his*⁺ | *try*⁺ | *his*⁻ | 28·6 | |
| 1b | *try*⁺ | *his*⁻ | *try*⁻ | *his*⁺ | 19·0 | 23·8 |
| 2a | *his*⁻ | *ant*⁺ | *his*⁺ | *ant*⁻ | 45·0 | |
| 2b | *his*⁺ | *ant*⁻ | *his*⁻ | *ant*⁺ | 67·4 | 56·2 |
| 3 | *try*⁻ | *ant*⁺ | *try*⁺ | *ant*⁻ | 43·5 | — |

the only satisfactory way of obtaining information on the order of genes is by 3-factor crosses (see below).

In the above example, the mutational sites under consideration affected different aspects of the phenotype and so mutations in one functional region (e.g. *try*) could be easily distinguished from those in another (e.g. *ant*). But, clearly, when the two mutations determine a similar change in the phenotype they cannot be distinguished except by breeding. There are two methods which can be adopted under these circumstances. In one, the frequency of wild-type transformants produced in a mutant$_1$ by mutant$_2$ experiment is expressed as a fraction of that obtained when a wild-type donor is used with a recipient which is mutant in one way or the other (Table 5.4). This method, therefore, involves a comparison of values obtained in different experiments and this is never wholly satisfactory.

The second method involves the introduction of a third marker and it allows the recombination frequency in the intercept under consideration to be expressed in terms of the frequency in a second intercept. Further, these two values can be determined from the outcome of the same experiment and so errors owing to extraneous factors are avoided. Thus, in order to study recombination in one intercept (limited by two mutations), a second intercept is defined by the introduction of a third marker. In effect, therefore, this is a 3-factor cross and, as intimated above, crosses of this kind yield information concerning the order of the loci involved in it. The basis of the method is as follows. Let us suppose that preliminary 2-factor crosses have shown that loci *B* and *C* both lie on the same side of *A* but the alternative sequences *ABC* and *ACB* have

**Table 5.4** Mapping mutant sites by comparing the frequency of wild-type recombinants obtained in mutant$_1$ donor × mutant$_2$ recipients with those obtained in wild-type donor × mutant$_2$ recipients. The symbols $x$, $y$ and $z$ represent the exchange frequencies in their respective intercepts. By expressing the frequency of wild-type recombinants in Cross I as a fraction of those in Cross II, the recombination frequency in the $m_1$–$m_2$ intercept is essentially expressed as a fraction of that occurring in an adjacent intercept. Thus, $yz/z(y+x) = y/y+x$ from which $y/(y+x) - y$ can be obtained.

| Cross Donor | Cross Recipient | Minimum exchanges required for production of wild-type | Expected frequency |
|---|---|---|---|
| $m_1$ | $m_2$ | $\begin{array}{cc} x & y & z \\ m_1 & m_2^+ \end{array}$  — $\times$ —— $\times$ —  /  $m_1^+$   $m_2$ | $yz$ |
| + | $m_2$ | $\begin{array}{cc} m_1^+ & m_2^+ \end{array}$  — $\times$ — $\times$ —  /  $m_1^+$   $m_2$ | $yz$ |
|  |  | or | + |
|  |  | $\begin{array}{cc} m_1^+ & m_2^+ \end{array}$  $\times$ ———————— $\times$  /  $m_1^+$   $m_2$ | $xz$ |
|  |  |  | $= z(y+x)$ |

not been distinguished. Reciprocal 'matings' of the type $ABc \times abC$ and $abC \times ABc$ are then set up and the argument proceeds as follows:

1. If the order is $ABC$:

   (a) In $ABc$ (recipient) × $abC$ (donor) crosses, the production of wild-type products requires a minimum of two recombinations.

   (b) A minimum of two recombinations is required for their production in the reciprocal cross also.

2. If the order is $ACB$:

   (a) Two exchanges is still the minimum number in the case of $AcB$ (recipient) × $aCb$ (donor) matings, but

(b) A minimum of four exchanges is required in the reciprocal cross.

Thus, a comparison of the frequency of wild-type products in reciprocal crosses enables the sequence to be determined (Table 5.5).

**Table 5.5** The effect of gene sequence on the frequency of wild-type recombinants produced following reciprocal *ABc* × *abC* crosses.

| Mating | | Exchanges required to produce wild-type (*ABC*) recombinants | |
|---|---|---|---|
| Donor | Recipient | With *ABC* sequence | With *ACB* sequence |
| *abC* | *ABc* | a b C / X X / A B c | a C b / X X / A c B |
| *ABc* | *abC* | A B c / X    X / a b C | A c B / X X X X / a C b |
| Comparison of crosses | | Two exchanges required irrespective of direction of cross | Two exchanges in one cross and four in its reciprocal |

As indicated earlier, accurate mapping by transformation is feasible only for short segments at a time because DNA molecules in the transforming principle each represent only about a half of one per cent of the bacterial genome. But the segments separately mapped by the methods outlined above can be arranged relative to each other by the method described on p. 30. A technically very different method from this, but one having a similar basis and purpose in this connection, can be used for long-range mapping by transformation. It depends on the difference in buoyant density of $^{14}N$ and $^{15}N$ polynucleotide columns and their separability by density gradient centrifugation. In outline, the method is as follows. Non-dividing bacteria whose DNA is uniformly labelled with $^{15}N$ are transferred to a medium containing $^{14}N$. DNA is then extracted from these bacteria at intervals within the period of the first replication cycle following transfer. The extracted DNA is centrifuged to separate the hybrid $^{14}N$-$N^{15}$ DNA from the unreplicated $^{15}N$-$N^{15}$ DNA and the hybrid DNA alone is used in transformation experiments. Now, since replication is initiated at a fixed point and proceeds in one physical direction from that point (p. 29), the early samples of hybrid DNA will be able to effect the transformation of only those character-differences controlled by genes near the point of replication initiation. Subsequent samples will be able to transform progressively more characters and, clearly, the progression will reflect position:time marks location.

## Mapping by transduction[25]

In transformation, DNA is irreversibly taken up very quickly by the bacterial cells. Thus, transformants are produced even if DNase is added as little as 10 seconds after the bacteria are exposed to transforming DNA. Further, this DNA is incorporated into the recipient genome very quickly and well before the recipient cells divide. The evidence shows also that the donor DNA is not incorporated as a supernumerary segment. On the contrary the incorporated material replaces homologous sections of the resident genome. Thus, transformed nuclei are purely haploid and transformed strains are as stable as the original forms.[61]

Transductants, on the other hand, are not entirely stable even when complete transduction is involved (p. 62). For example, the transducing phage P1 establishes lysogeny before the recipient bacterium divides but several cell generations may elapse prior to lysogenization by P22. In its case, therefore, a period of 'abortive transduction' may precede 'complete transduction'. This, of course, implies a temporary period of partial diploidy. In fact it would seem that merozygosity is the customary condition following complete transduction also, at least in the case of restricted transduction. For example, galactose negative ($gal^-$) strains of *E. coli*, transduced to the $gal^+$ state by lambda phage, produce $gal^-$ haploid segregants at low frequency (c. $10^{-3}$ per cell per generation). These segregants are as stable as the original recipients and, if transduction arose originally from a single infection, they are no longer immune to lytic super-infection. This indicates that the transducing DNA need not replace any part of the recipient genome and the transductant in the above example can be symbolized as $gal^-/gal^+$. Merozygotes of this kind are called heterogenotes, the resident genome being called the endogenote and the donor fragment the exogenote. This condition of partial hybridity provides a basis for recombination and thus for mapping also.

Many complications arise in regard to the detection of linkage in transformation experiments but these do not obtain in the case of transduction.

Thus, in LFT lysates, only about one phage particle in $10^5$–$10^7$ carries a particular donor marker. Consequently, even if each bacterium absorbs 10 particles, double transductants for unlinked markers are expected with a frequency of only about $10^{-8}$ which approximates to the mutation rate for single markers. However, the observed frequency of joint trans-duction of markers known on other grounds to be linked is about a thousand times higher than this.

The locations of prophages responsible for restricted transduction can be mapped by the gradients of their transfer and transmission during conjugal bacterial mating (see p. 58). In fact the consequences of con-jugation between bacteria which are lysogenic for different strains of the same phage have shown that recombination can occur between prophages which maintain their association with the bacterial chromosome.

What is more, transducing phage can carry not only bacterial genes but other prophages. For example, prophage 18 was first shown to be located in the vicinity of two loci concerned with methionine synthesis in *E. coli* by the fact that it and these markers could be jointly transduced by phage 363 which effects generalized transduction.

The question of pre- versus post-zygotic exclusion arises in all the methods used for mapping the bacterial chromosome. In the present context, the frequency of transfer can be judged from the number of abortive transductions. This frequency varies widely for unlinked markers but tends to be the same for linked genes. However, the fre-quency of complete transductants even for linked markers may show considerable variation depending on the selection procedures used for their isolation.

## Mapping by conjugation

Bacterial conjugation consists of a number of more or less distinguish-able phases. The process is initiated by random collisions which, when they occur between donor and recipient, are quickly converted into unions. Union is established by the formation of a linking conjugation tube. This implies a specificity the basis of which resides in the donor wall and for which the sex factor is responsible.

As in the case of transduction and transformation, the results of con-jugation are usually studied via the recombinants derived from recipient cells. For example, the kinetics of union formation and various other properties of bacterial conjugation can be investigated in the following way.

Suitably marked $F^-$ recipients and Hfr donors of a given strain are mixed in liquid culture at concentrations which favour collisions and the initiation of union. Samples of the mating mixture are then taken at intervals and diluted very carefully. Dilution serves to prevent further

collision and subsequent union but it is performed gently so that the
unions which have already occurred are not disrupted. The interval
available for union formation can thus be limited. The diluted mixtures
are then plated on selective media. The matings which were initiated
prior to dilution are completed on the plates and the outcome of the
mating can be studied. Clearly, the selective media must prohibit the
growth of pure donor genotypes. This can be achieved by making
the original donor susceptible to some agent (e.g. streptomycin) which
the recipients resist. Jacob and Wollman performed such an experiment
using the HfrH strain as a donor. This carried such markers as the
ability to synthesize threonine, leucine and tryptophan and to ferment
galactose, all of which the F⁻ recipient strain lacked. However, the
recipient was resistant to streptomycin while the donor cells were
susceptible, so pure donors could be inhibited easily. Recombinants for
each of the donor markers were selected separately by the use of suitable
media. Idealized results of such an experiment are shown in Fig. 5.4
from which the following facts emerge.

First, all the curves arise very close to the origin, which shows that

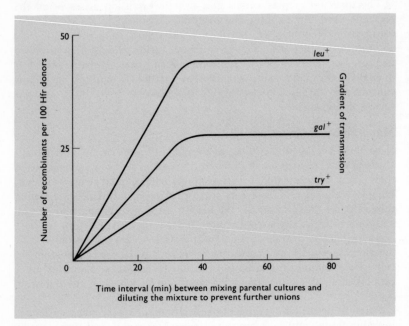

**Fig. 5.4** Idealized results of an uninterrupted mating experiment between HfrH
*leu⁺ gal⁺ try⁺* str-s donors and F⁻ *leu⁻ gal⁻ try⁻* str-r recipients when donor
markers are selected for independently.

union is initiated very quickly after mixing under the conditions of the experiment. Second, with respect to the recombinant yield, all the markers show a plateau at about the same time so that prolonging the time available for mating initiation beyond 30–40 minutes does not result in any further increase in the number of recombinants produced. Third, the level of the plateau varies with the marker but under carefully controlled conditions it is constant for a given marker in different experiments involving the same Hfr donor strain. Fourth, since all the curves begin very close to the origin, and reach a different plateau at the same time, the slope of curve varies with the marker also. Thus the markers can be arranged in a sequence according to the level of the plateau they reach. This arrangement reflects what is called the gradient of transmission.

If a single marker is considered on its own, the slope of the curve of transmission appears to indicate the rate of increase of union initiation as the time available for collision is lengthened. But clearly this does not explain the differences between the slopes of the curves for different markers. These must depend on events in the mating process which occur after the unions are established. These are of two main kinds: first, the transfer of the donor chromosome and, second, recombination in the merozygote.

Information on the transfer of the donor chromosome is provided by the now famous interrupted mating technique of Jacob and Wollman.[43] This can be illustrated by reference to the strains described in the above experiment where every precaution was taken not to interrupt the mating. In an interrupted mating experiment, the mates are separated, by agitation in a high speed blendor, at intervals after the initial mixing of donor and recipient strains. Selection for recombinants can then be achieved by plating on suitably selective media.

In the case in point, no recombinants of any kind were observed when mating was interrupted within about eight minutes of mixing. But when each marker is selected for independently, $leu^+$ recombinants are produced by the recipients which are allowed to mate for longer than eight minutes. In the case of $gal^+$, however, no recombinants arise unless mating is allowed to last for at least 22 minutes or so, while over 30 minutes of mating is necessary in the case of the $try^+$ marker.

Thus, when mating is interrupted, the transmission curves for the different markers do not have the same origin. But they eventually reach the plateaux observed in uninterrupted mating experiments and take about the same time to do so (Fig. 5.5). The sequence based on the order in which the markers make their first appearance is known as the gradient of transfer and it corresponds with the gradient of transmission: markers which appear earlier reach a higher plateau (cf. Figs. 5.5 and 5.6).

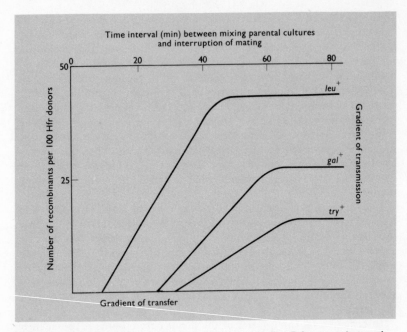

**Fig. 5.5** Idealized results of a cross similar to that in Fig. 5.4 except that mating was interrupted by blending. Note that the curves for all these markers are the same as their counterparts in Fig. 5.4 except that they have different origins on the time axis.

Instead of selecting for each marker separately, a recombinant population carrying a 'selected' marker can first be isolated and the frequency of the other 'unselected' markers in this population can be determined subsequently (p. 87). Now, when the selected marker is that which gives the highest plateau and appears earliest, the frequencies of the non-selected markers form a series which parallels those in respect of both transfer and transmission. Thus, $gal^+$ recombinants are more common than $try^+$ recombinants among those selected for the $leu^+$ marker. This relationship has been confirmed using many genes and it indicates that both the gradient of transfer and the subsequent gradient of transmission reflect the linkage relationships of the genes concerned. In other words, the results are consistent with the view that an Hfr donor transfers its chromosome as an oriented linear structure with a leading end (O for origin) and a trailing end. Sex factor functions are the last to be transferred and so a part, at least, of the sex factor must occupy the tail end. This accounts for the low rate of recipient to donor conversions in $F^- \times$ Hfr matings.

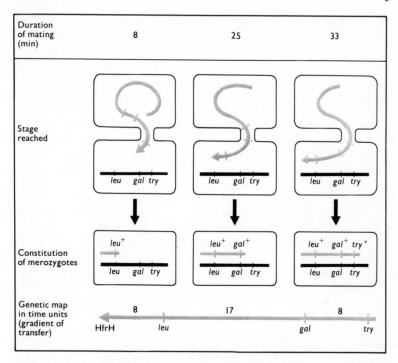

**Fig. 5.6**  Diagram illustrating the stage of genoneme transfer reached by the most advanced mating pairs 8, 25 and 33 minutes after mixing the parental cultures of the experiment illustrated in Fig. 5.5. For convenience only the proximal parts of the genonemes are drawn and both are shown as linear.

Thus, marker genes can be arranged in sequence on the basis of two mutually consistent criteria, namely, the gradient of transfer and the gradient of transmission. But what is the relationship between them, and how is the time unit of the former related to the percentage frequency of the latter? Two further questions must be considered before these can be answered.

Earlier it was pointed out that various difficulties arise in trying to distinguish between pre- and post-zygotic influences (see p. 90). In the present connection we must attempt to determine the extent to which the gradient of transmission depends on the gradient of transfer, that is, on pre-zygotic exclusion. It is conceivable that donor genes which arrive early have a greater chance of being transmitted to recombinants than donor genes which arrive late. On this basis a gradient of transmission would obtain even if a complete donor genome was always transferred during conjugation (post-zygotic exclusion) but the markers could still be arranged in a physical sequence. Alternatively, all markers once received may have an equal opportunity of being transmitted to recombinant progeny in which case the gradient of transmission would depend entirely on pre-zygotic exclusion. The experiments which resolved this issue rest on a phenomenon called zygotic induction which must be described before the experiments can be considered (p. 59).

The outcome of conjugal crosses between lysogenic and non-lysogenic bacteria depends on the direction in which the cross is made. Thus, if for example, the F$^-$ strain is non-lysogenic for lambda phage, mating is often followed by the initiation of a lytic cycle in the recipient. Interrupted mating experiments show that this so-called zygotic induction does not occur unless mating is maintained for the length of time normally required for the transfer of the *gal* region. Now, lambda can effect the localized transduction of the *gal* region and the results of zygotic induction in interrupted mating experiments support the chromosome location of lambda prophage. Thus, the frequency of zygotic induction following F$^-$lys$^-$ × Hfr lys$^+$ mating is a measure of the frequency of transfer of the lambda prophage location; recombination is not involved because the phage is expressed in the zygote itself. In view of the close linkage between lambda and *gal*, the transfer frequency of the two must be approximately equal. The plateau of zygotic induction represents the frequency of transfer and, in this case, is equivalent to the frequency of transmission also. Now, the plateau for the gradient of transmission of *gal* is only about a half of that for lambda when, as we have reasoned above, their frequency of transfer is about the same. This must mean that a donor gene transferred to a recipient has a fifty-fifty chance of being transmitted to the recombinant progeny. In other words it has as much chance of being transmitted as the homologous region resident in the recipient. This chance is known as the coefficient of integration.

The question we posed earlier was this—is the coefficient of integration the same for all the markers received or does it depend on the time of entry? Now phage 21 transduces the *try* region which, as we have seen, is transferred later than *gal* in F$^-$ × HfrH matings. But in this case also, although the plateaux for the frequency of transmission are lower for both the phage and the locus it can transduce, the frequency of phage transfer and that for gene transmission have a 2:1 relationship. Thus, the coefficient of integration is not affected by the time of entry and the gradient of transmission depends entirely on the gradient of transfer (i.e. pre-zygotic exclusion).

This means that maps based on the gradient of transmission do not depend on variations in recombination frequencies. Even so, as we intimated earlier, they are perfectly valid maps in regard to order and with regard to relative map distances also, provided that the breaks which determine pre-zygotic exclusion are distributed at random.

Uninterrupted mating experiments with variously located temperate phage capable of effecting localized transduction have shown that the frequency of zygotic induction falls exponentially as the interval between the origin and the transfer time of the prophage location increases. This favours the idea that there is a random distribution of breakage. Further, the frequency of breakage can be related to length provided that the chromosome moves at constant rate during transfer. The leading extremity (O) varies from one independently-isolated Hfr strain to another. Consequently, while mutually consistent maps can be constructed (see p. 109), individual markers are transferred at different times by different Hfr donors. The sequence also may be different. However, under carefully controlled conditions, the interval between the times at which two linked loci are transferred is the same for different Hfr strains unless, of course, the origin falls between the two loci in some cases and outside them in others. Within these categories, however, the above constancy obtains. This indicates that the chromosomes transfer occurs at constant rate.

To recapitulate:

1. The marker sequence in the gradient of transmission corresponds with that in the gradient of transfer and both reflect the linear order of the genes on the genoneme.
2. The coefficient of integration is the same for all markers irrespective of their time of entry so that the gradient of transmission is determined by the gradient of transfer, and
3. Exclusion during oriented transfer depends on breakage which is randomly distributed in relation to length.

From these it is clear that the time units in which the gradient of transfer is measured correspond to length units in the physical chromosome and both are related to the unit in which the frequency of transmission is measured. But before considering the 'constants' which permit these various units to be interconverted, a further unit of mapping must needs be considered, namely, the unit of recombination which was used in the previous techniques described.

Clearly, recombination, uncomplicated by other factors such as pre-zygotic exclusion which also affect recombinant frequency, can be studied only if the zygotes concerned are uniform in their constitution. In other words, only those zygotes which receive all the marker genes under consideration should be taken into account. In conjugation experiments this is easily achieved by selecting recombinants which carry the marker gene which is furthest from the origin. Clearly, in view of the orientated transfer, these must be descended from zygotes which also received all the other markers in the intercept between the origin and the selected marker. The frequency of these unselected markers among the selected recombinants can then be determined and a map based on the re-combination values can be constructed (see p. 104).

Thus, the relationship between genes can be expressed in terms of:

1. The time interval which elapses between their transfer. This is revealed in the gradient of transfer as determined in interrupted mating experiments.
2. The difference between the plateaux observed in relation to their transmission to recombinants. These are revealed in both interrupted and uninterrupted mating experiments whether selected markers are used or not.
3. The recombination values between them. These are determined from the linkage relationships found in the products of a homogeneous population of merozygotes.

The various units employed in these determinations can be equated not only with each other but with physical distance. Indeed, they can be expressed in terms of the molecular structure of DNA (Table 5.6).

**Table 5.6** Approximate equivalence of the units in which the genoneme of *E. coli* can be measured.

| | UNITS | | | |
|---|---|---|---|---|
| | Physical length ($\mu$) | Transfer time (min) | Recombination units | No. of base pairs |
| Total | 1100 | 90 | 1800 | $3 \cdot 15 \times 10^6$ |
| Equivalence | 1 | 0·09 | 1·67 | $2 \cdot 9 \times 10^3$ |
| | 12 | 1 | 20 | $3 \cdot 5 \times 10^4$ |
| | 0·6 | 0·05 | 1 | 1750 |

As indicated above, loci which are more than about two time units apart appear to be unlinked in recombination studies. Therefore these analyses can be used only for the mapping of regions which are less than two time units long. In fact, when the distance between loci is less than two minutes, mapping by the gradients of transfer and transmission is not sufficiently accurate, and recombination analysis must be employed. The gradients of transfer and transmission must be used, however, for long distance mapping. But a given Hfr strain transfers only about a third to a half of its chromosome with high frequency and so various donor strains are required to map adequately the whole bacterial genome. In fact it was the use of a variety of Hfr donors which revealed the circular nature of the linkage map in *E. coli*. For example, if we again consider the marker genes *leu*, *gal* and *try* discussed earlier, the following gradients of transfer and transmission are shown by various independently-isolated Hfr strains:

| | | | |
|---|---|---|---|
| Hfr(H) | *leu* | *gal* | *try* |
| Hfr(B8) | *gal* | *try* | *leu* |
| Hfr(J4) | *try* | *gal* | *leu* |
| Hfr(C) | *leu* | *try* | *gal* |

These are consistent with a circular structure which can break at various places prior to transfer, the break-point and, thus, the orientation of transfer, being constant for a particular Hfr strain.

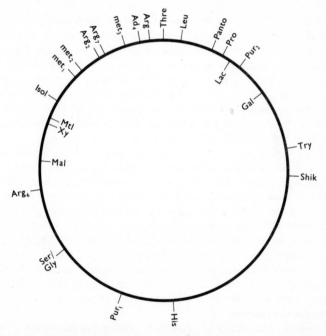

**Fig. 5.7** The circular linkage map of *E. coli*. Symbols on the outside refer to the capacity to synthesize purines and various amino acids, those on the inside to various fermentation capacities. Only a small sample of the known sites is given. Hfr and phage locations are shown in Fig. 4.2. Note the rarity of known sites in the so-called 'silent region' (second quarter).

To sum up:

Mapping in bacteria can be performed on the basis of transformation, transduction or conjugation and, in the last, the frequency of transfer, transmission or recombination can be employed. Some bacteria are more amenable to analysis by a particular method. Further, the techniques themselves are to some extent complementary because some are suitable only for short range mapping while others are preferable, or even necessary, for mapping over long distances.

Each method presents its own operational and interpretative problems but on the credit side is the high resolution achieved in these organisms which can be grown quickly and selectively in large numbers. The high resolution of the methods is amply demonstrated by the extent to which the millimetre of DNA in *E. coli* has been mapped (Fig. 5.7).

## MAPPING IN STANDARD SEXUAL SYSTEMS

Linkage groups in higher organisms can be represented as maps on which the various linked loci are distributed according to their recombination behaviour relative to each other. These maps are usually linear and the order of the genes on the map generally corresponds with their sequence on the chromosomes. Within limits, the relative distances between genes on these maps are equivalent to their spatial distribution on the chromosome. But before the techniques employed in making these maps and the interpretative problems they present are described, the rationale of the methods will be considered.

Both the bivalents at the first meiotic division and the chromosomes at the second meiotic division orient at random relative to each other. Consequently, genetic differences on non-homologous chromosomes recombine at random regardless of the occurrence, frequency or distribution of crossing-over. Thus, if two gametes $AB$ and $ab$ fuse to give a zygote $AB/ab$ then, following meiosis, the original parental combinations, $AB$ and $ab$, and the new, recombined types, $Ab$ and $aB$, will be produced in equal numbers. Likewise, the reciprocal cross $(Ab \times aB = Ab/aB)$ will produce a zygote which, on meiosis, yields equal numbers of the two parental-type, $Ab$ and $aB$, and the two recombined type, $AB$ and $ab$, gametes. This state of affairs is known as independent assortment or random recombination.

But clearly, irrespective of orientation, two genes on a given chromosome will be transmitted together unless crossing-over separates them. In this event, in so far as they are separable by crossing-over at all, the two genes will show non-independent assortment or partial linkage or, at least, they will show linkage via other genes in the same linkage group.

Thus, with partial linkage, zygotes of the type *AB/ab* are expected to produce a preponderance of the parental-type gametes, *AB* and *ab*, and a deficiency of recombined-type gametes, *Ab* and *aB*. Of course, the two parental types, on the one hand, and the two recombined types on the other, are expected with equal frequency because chromosome reproduction is accurate and crossing-over is generally reciprocal (but see p. 126). What is more, where linkage obtains, differences are expected between the gametic frequencies of comparable zygotes produced by reciprocal crosses. Thus, the reciprocal zygote *Ab/aB* produced by the fusion of *Ab* and *aB* gametes is also expected to produce a preponderance of parental-type gametes but in this case these are *Ab* and *aB*. In both cases, however, the frequency of recombined types is less than 50%.

The most generally applicable method of mapping in sexual organisms is based on scoring a random sample of the gametes produced by a particular heterozygous diploid type. The vast majority of the mutations which arise in nature or experiment have no obvious effect on the phenotype of the gametes themselves. Consequently, their genetic nature can be determined only by examining the vegetative phase to which they give rise, either individually as in haploid cycles, or following their fusion in essentially diploid cycles.

The frequency of recombined gametes obtained is tested to see whether it departs significantly from that expected on the basis of random assortment. If the departure is significant and other sources of deviation (e.g. differential viability) can be excluded then the loci are held to be linked.

The degree or extent of linkage is then expressed simply in terms of the percentage frequency of recombined-type gametes. This is called the recombination value. Thus, the recombination value for loosely linked loci is expected to approach 50% while that for closely linked genes will approach zero.

In the absence of intervening markers, two loci are placed on the linkage map at a distance apart equal to the recombination value between them. For example, if the joint frequency of *Ab* and *aB* gametes from an *AB/ab* diploid is 10%, the two loci are placed 10 units apart on the linkage map. Similarly if an *AC/ac* zygote gives *Ac* and *aC* gametes with a joint frequency of 15%, the alpha and gamma loci are placed 15 units apart. These two results show that the gamma locus is further from the alpha than is the beta locus. But they do not allow preference to be expressed between the sequences *ABC* and *CAB*. In fact, the order can be determined only if the recombination value in respect of beta and gamma is known also. Generally, and, in most cases, preferably, all three recombination values are estimated from the results of one experiment in which all three loci are segregating (e.g. *ABC/abc*). Such a cross is called a three-point experiment.

Linkage maps constructed in this way serve two immediate functions. First, they show the linear order of the mutant sites, at least in organisms with a normal meiotic sequence. Second, they give some indication of the amount of recombination to which a chromosome region is subjected and, if exchange events are distributed randomly along the chromosome, they indicate the spatial distribution of the marker loci. But what is the relationship between the recombination value in respect of two loci and the total amount of crossing-over per chromatid (cross-over value) in the intercept between them? With regard to the production of nuclei which are recombinant in respect of any two particular loci, cells undergoing meiosis can be classified into two main groups:

1. Cells in which no exchanges at all occur between the two loci concerned, and
2. Cells in which one or more recombinational events do occur in the segment defined by the marker loci.

Clearly, the former can give rise only to parental-type gametes so far as the marker loci are concerned. But, excluding sister-strand exchange and chromatid interference, the second group will produce equal numbers of parental-type and recombined-type gametes in respect of the markers. This holds irrespective of the number of exchanges which occur in the intercept between them (Fig. 5.8). In other words, if the second group (i.e. those with one, two or more exchanges in the segment concerned) represent $z\%$ of the cells undergoing meiosis, $0.5z\%$ of the products will be recombinant for the marker loci. This means that the directly-observable recombination value for these loci will be $0.5z$. If, therefore, these loci are plotted without reference to intervening markers, they will appear $0.5z$ units apart on the linkage map.

From this it should be clear that the recombination value does not reflect accurately the extent of exchange in an intercept because the recombination value is directly related to the average frequency of cells with crossing-over rather than the average frequency of crossing-over in the cells. The recombination value and the cross-over value are equal only when the cells with recombination never have more than one exchange in the intercept concerned and this situation is approached only when very short segments are considered.

What, then, is the general relationship between the average frequency of exchange in a segment and the frequency of cells with exchange in that segment? On the assumption that the average frequency of recombination in the intercept is comparatively low and that the various exchanges occur independently of one another, this relationship can be calculated on the basis of a Poisson distribution.

Thus, where $m$ equals the average frequency of chiasma formation in

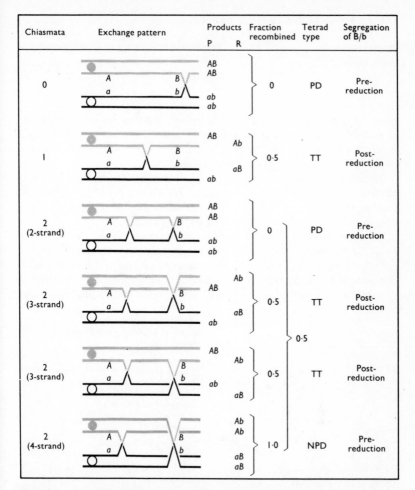

**Fig. 5.8** The relationship between chiasma formation and recombination. Two loci on the same chromosome are recombined only when exchange occurs between them. Each exchange involves only two of the four strands. Consequently the exchange frequency per strand is only half the chiasma frequency. But the frequency of recombinants in respect of any two loci is equal to half the frequency of cells with chiasmata between them. The number of chiasmata in the intercept is irrelevant because on average 2 parental and 2 recombined gametes arise whether 1, 2 or n chiasmata occur in it. This conclusion ignores sister strand crossing-over and chromatid interference.

PD = Parental ditype
NPD = Non-parental ditype
TT = Tetratype

Note that the frequency of post-reduction is equivalent to the frequency of tetratype tetrads (cf. Figs. 5.9 and 5.11 and see p. 119).

an intercept, the probability of $n$ chiasmata occurring in that intercept is given by:

$$x_n = \frac{e^{-m}m^n}{n!}$$

Since the factorial of o is equal to 1, and any number raised to the power of nought is also equal to one, both $n!$ and $m^n$ are equal to one when $n = 0$. In other words, the frequency of cells in which no chiasmata occur in the critical intercept ($n = 0$) is equal to $e^{-m}$.

Therefore, by subtraction, the probability of obtaining cells with one or more recombinations in the intercept is equal to $1 - e^{-m}$. And the recombination value is half this amount because only two of the four strands in a bivalent are involved in any one chiasma.

When $m$ is small, $1 - e^{-m}$ is very nearly equal to $m$. But although $1 - e^{-m}$ increases with the value of $m$, it does so at a decreasing rate and tends to a limit (Fig. 5.9). This is because the mode of the distribution curve approaches the mean as the value of $m$ increases (and the distribution tends towards normality). And while cells with many chiasmata

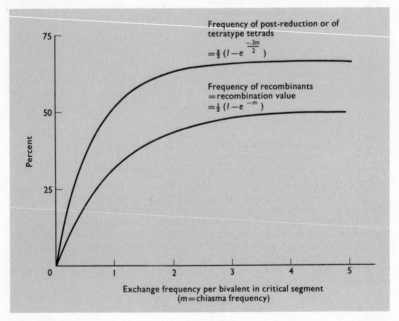

**Fig. 5.9** The relationship between mean chiasma frequency, frequency of post reduction ( = tetratype frequency) and the recombination value (cf. Figs. 5.8 and 5.11).

contribute proportionally more than the single-chiasma cells to the magnitude of the mean chiasma frequency, they contribute only as much as the single-chiasma cells to the recombination value. Thus, while $m$ has no theoretical upper limit, the recombination value cannot exceed 50% for, clearly, $1 - e^{-m}$ cannot be greater than unity.

Since the cross-over value per chromatid is directly related to the chiasma frequency per bivalent, the maximum length of the linkage map is limited by the chiasma frequency. The extent to which the observed map length approaches this limit is a reflection of the intensity of the mapping. For example, if the number and distribution of available markers are such that all the exchanges which occur can be detected, the map will reach the limit of its length. Increasing further the density of markers will not lead to any further increase in map length because the extra markers will simply bisect segments within which the recombination values are additive anyway (i.e. the recombination values equal the cross-over values). Thus, on the assumption that all chiasmata represent exchanges and all exchanges are revealed as chiasmata, the limit of map length can be calculated (p. 217).

In summary, the recombination value and the cross-over value per chromatid are very nearly equal to each other and to half the mean chiasma frequency when this is low. The chiasma frequency is low when the segments under consideration are short. Under these circumstances the recombination values of adjacent segments are very nearly additive but over longer lengths they become increasingly less so.

Thus, when long segments are mapped on the basis of the two loci which define them, they appear to be shorter than they are when the map distance is determined by adding the recombination values for intervening regions. For example, if we consider a segment A-E marked by uniformly distributed intervening loci B, C and D so that the recombination value for each intercept is 10, then the map distance for A-E obtained by addition is forty. But if the intervening loci are ignored, the directly observed recombination value of A and E will be only three-quarters of this. Thus, if we assume that crossing-over occurs independently in each of the intervening intercepts, the percentage of cells without exchanges between A and E is expected to equal $(0.8)^4$. Therefore, those with an exchange in this segment should equal $1 - (0.8)^4$ and, for reasons given earlier, the observed recombination value will be half of this, viz. 29.5%. In fact, if a segment is long enough to contain more than one chiasma it can be mapped accurately only if intervening loci are considered and both the number and distribution of the intervening markers affect the accuracy of the mapping (see Table 5.7).

A number of assumptions were made in the above discussions. In particular it was assumed that the distribution of chiasmata was at

**Table 5.7** The relationship between map distance and the number and distribution of intervening markers. Recombination between segments is assumed to be independent and the recombination value within each of the smallest segments is assumed to be ten.

| Number of intervening markers | A —\|————— B —\|————— C —\|————— D —\|————— E —\|– | Map distance A–E |
|---|---|---|
| | 10    10    10    10 | |
| 3 | ←—10—→ ←—10—→ ←—10—→ ←—10—→ | 40 |
| 2 | ←—10—→ ←—10—→ ←————18————→ | 38 |
| 1 | ←————18————→ ←————18————→ | 36 |
| 1 | ←—10—→ ←——————24·4——————→ | 34·4 |
| 0 | ←———————————29·5———————————→ | 29·5 |

random. However, cytological examination shows that although chiasmata appear to be more or less randomly distributed along the chromosome in many cases, their extreme localization in particular regions is not uncommon. This means that a segment from a region of high chiasma frequency will occupy a longer length of the linkage map than a segment of equal physical length from a region where chiasma formation is rare or uncommon. What is more, breeding experiments have shown that, when intergenic recombination in short segments is considered, the frequency of double, triple, etc. crossing-over is less than that expected on the basis of independent events. This phenomenon is known as positive interference.

Consider, for example, a segment ABC where the recombination value for both A-B and B-C is 10%. On the assumption that crossing-over in one intercept is independent of that in the other, coincident crossing-over in both segments is expected in 10% of 10% = 1% of the cases. The following values are then expected in a three-point cross:

| Gametic types | Parental | Single cross-overs | | Double cross-overs $A \times B \times C$ |
|---|---|---|---|---|
| | | $A \times BC$ | $AB \times C$ | |
| | ABC and abc | Abc and aBC | ABc and abC | AbC and aBc |
| Percentage frequency | 81 | 9 | 9 | 1 |

If, at the other extreme, we assume that crossing-over in one intercept so interferes with that in an adjacent intercept as to make the two events mutually exclusive, then, clearly, no double cross-overs are expected at all. In this event a three-point cross is expected to give the following results:

| Gametic types | Parental | Single cross-overs | | Double cross-overs |
| | | $A \times BC$ | $AB \times C$ | $A \times B \times C$ |
| --- | --- | --- | --- | --- |
| | ABC and abc | Abc and aBC | ABc and abC | AbC and aBc |
| Percentage frequency | 80 | 10 | 10 | 0 |

Notice that both these results give the same recombination value of 10% for *AB* and for *BC* and the same map would be produced on the basis of them. But what happens in practice?

If the map lengths involved are long, the results accord closely with those expected on the basis of non-interference but when the lengths are short the results approach those of the second expectation. In the latter event, the recombination values for adjacent segments tend to be additive and this situation has been found to hold for segments less than 10 map units long in *Drosophila*.

The extent and direction of interference is usually expressed as the coefficient of coincidence which is simply the extent to which the number of double cross-overs observed conforms with that expected on the basis of independent events. Thus:

Coefficient of coincidence $= c$

$$= \frac{\text{Observed number of double cross-overs}}{\text{Expected number of double cross-overs}}$$

In the first situation considered above, the observed number (by design) equalled the expected number so that coincidence was equal to 1. In the second situation, again by design, there were no double cross-overs and so the value of *c* was nought. Intermediate values indicate degrees of positive interference.

The occurrence of such interference means that there are fewer multiple cross-overs than expected on the basis of random events. Consequently, the actual recombination value does not fall off quite as quickly with increasing chiasma frequency as the assumption of random-ness leads one to expect. In other words, since doubles tend to be less frequent because of interference, the relationship between the mean exchange frequency and the frequency of cells with exchange is closer than that predicted on a Poisson distribution.

To what may interference be attributed? It is clear that the minimum requirements for the production of a doubly recombined strand are:

1. The occurrence of two exchanges in the bivalent from whence it came, and
2. The involvement of a particular chromatid in both these exchanges.

Consequently, the observed frequency of double recombinants will depart from expectation if:

1. The occurrence of one chiasma interferes with the prospects of another chiasma being formed in its vicinity. This so-called chiasma interference will affect the frequency distribution of chiasmata.
2. The chromatids involved in one chiasma are not selected at random in relation to those which participate in another chiasma (chromatid interference).

The two types of interference have similar consequences and they can be distinguished only under special circumstances. For example, in cases where chromatid interference is complete, so that two chromatids involved in one chiasma are prohibited from participating in a second, double recombinants cannot be produced. Under this circumstance the directly observed recombination frequency for AC will equal the sum of the recombination values obtained for the intervening intercepts AB and BC. And since the values for AB and BC could each exceed 25%, the directly observed value for AC could exceed 50%. But this situation cannot arise from chiasma interference.

Except for this unique effect, the two types of interference can generally be distinguished only if the products of a given meiotic cell can be analysed. This can be achieved in organisms where the products of a given meiosis remain together in a group. This condition is satisfied by many algae and fungi. The latter include cases where the meiotic products form what are called ordered tetrads. In these, the meiotic products are formed in single file and they usually retain the location in which they were formed. Consequently, the planes of the first and the second meiotic divisions can be distinguished and this allows the centromere position to be mapped.

From Fig. 5.10 it can be seen that if chiasma formation does not occur between a locus and the centromere, that locus will segregate at the first division of meiosis (pre-reduction). But if a chiasma does form, this segregation will not occur until the second division (post-reduction). Thus, the frequency of post-reduction is a measure of the exchange frequency. However, the relationship between the number of exchanges in the critical intercept and the frequency of post-reduction is rather

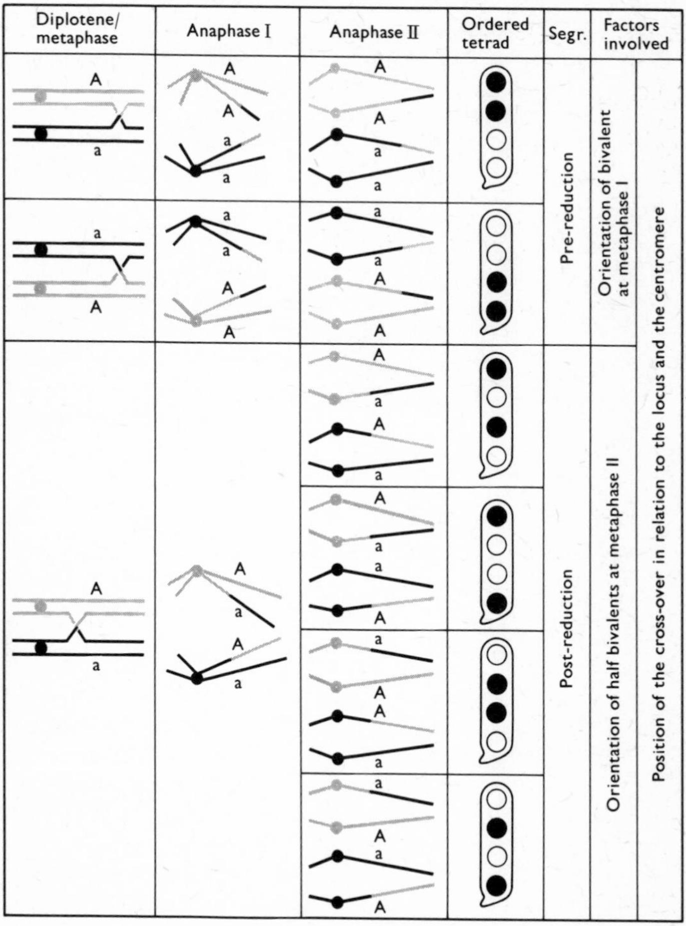

**Fig. 5.10** The effect of the location of crossing-over on the stage at which an allelic difference is qualitatively segregated. Pre-reduction: If the locus and the centromere are not recombined, the first division is reductional for that locus (AA–aa) and the second divisions are equational (A–A and a–a). Post-reduction: If the locus and the centromere are recombined, the first division is equational for that locus (Aa–aA) and the second divisions are reductional (A–a and a–A).

5

more complicated than that between the exchange frequency and the recombination value. Thus, as we have seen, 50% recombinants are produced whether one or more chiasmata occur between two loci. But while one exchange between the centromere and a locus gives 100% post-reduction for that locus, two exchanges (in the absence of chromatid interference) give only 50% while three exchanges give 75% post-reduction. In fact, the frequency of post-reduction is given by $\frac{2}{3}[1-(-\frac{1}{2})^n]$ where $n$ is the number of exchanges (Fig. 5.11). If this expression is combined with that for the Poisson distribution, the relationship between the frequency of post-reduction and the mean chiasma frequency $(m)$ is given by $\frac{2}{3}(1-e^{-\frac{3}{2}m})$. This function is plotted in Fig.

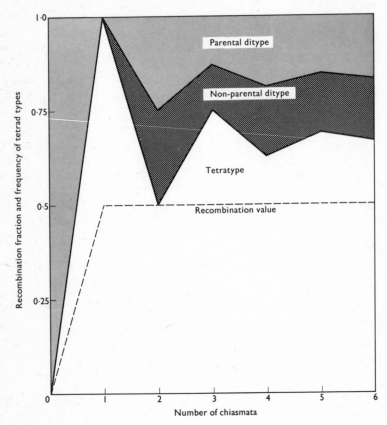

**Fig. 5.11**   The relationship between the number of chiasmata, the frequency of post-reduction (or tetratype tetrads) and the recombination value (cf. Figs. 5.8 and 5.9).

5.9 together with that relating the recombination fraction to the mean chiasma frequency. Thus, for small values of $m$ the frequency of post-reduction is very nearly equal to the mean chiasma frequency and to twice the recombination value. But for large values of $m$ the post-reduction frequency tends to a limit of two thirds while the limit for the recombination value is a half. In this event the frequency of post-reduction exceeds the corresponding recombination value by a factor of 0·75.

The frequency of post-reduction can be determined very easily when the gene difference concerned affects the appearance of the spores themselves. In other cases, the spores must be removed in sequence from the spore sac, their position of origin noted and their genotypes determined from the nature of the mycelia to which they give rise.

From Fig. 5.8 it is clear that 2-, 3- and 4-strand double exchanges have different consequence which can be distinguished if single tetrads can be analysed and, as indicated above, this can be achieved in certain organisms. Thus, the analysis of ordered tetrads can provide a great deal of information which is not obtained from a study of random samples. This will not be discussed in detail but some of the tetrad analyses which have contributed to knowledge of the mechanism of crossing-over are described later (see p. 123).


## MAPPING EXTRACHROMOSOMAL ELEMENTS

The genetic systems of extrachromosomal elements with genetic continuity are comparable in some ways with those of a non-integrated bacterial sex factor or vegetative virus. If, as in viruses, recombination occurs within, as well as between, homologous plasmagenes and plastogenes, a basis for mapping the genetic material of these elements is provided.

Sager and Ramanis have presented evidence in favour of recombination involving two unlinked non-chromosomal genes (acetate and streptomycin) in *Chlamydomonas*.[76] The site of action of these genes and their mode of recombination is unknown. Neither is it yet known whether specified extrachromosomal elements show recombination of the crossover type but a method for investigating this situation is being developed by Wilkie and Thomas. Many antibacterial antibiotics specifically inhibit protein synthesis in the mitochondria of yeasts. But a number of mutants which are resistant to these agents have been isolated. Mutants of this type show extrachromosomal transmission of resistance and suitable tests have indicated that the genetic elements involved are located in the mitochondria themselves. Many erythromycin-resistant mutants come into this category as do some which are resistant to one or other of various aminoglycoside antibiotics. Independently-isolated mutants of this kind provide a basis for mitochondrial genetic analysis because the mutations do not affect the respiratory capacity of the cells or their sexual fertility. Thus, in theory at least, studies of recombination, compensation and complementation are possible.

# 6

# The Mechanism of Recombination

## TETRAD ANALYSIS

Genetic analyses in higher plants and animals often involve extended chains of inference because the genetic nature of the meiotic products is rarely or barely revealed in the phenotype of the haplophase. Gametes and, therefore, the consequences of meiosis have to be studied through the zygotes produced by gametic fusion.

What is more the zygotes, the gametes and, hence, the meiotic cells of a given parent cannot but be treated as a population. Consequently, the outcome of a given meiosis can be defined, and the rules governing segregation and recombination can be formulated, only in statistical terms.

However, an examination of ordered, or, at least, individual, tetrads in fungi permits a more precise analysis of the consequence of individual meiotic divisions. In fact, tetrad analyses have led to a reconsideration of the statistical laws governing the sexual transmission of the genetic material. They have also forced a reconsideration of the mechanism of crossing-over and molecular models of this process have been proposed to accommodate the anomalous features which only tetrad analysis could reveal. The following case histories are typical of tetrad analyses in fungi, the differences between materials and experiments being ones of degree rather than kind.

### The *Sordaria* study

About ten distinct mutations affecting spore colour are known in the ascomycete *Sordaria fimicola*. Each of these is expected to show a normal

4:4 segregation in the eight-spored asci produced following crossing with the black-spored wild-type. Some of these mutations (e.g. $b_1$, $o$ and $st$-4) invariably show this expected behaviour. The others also behave normally in the vast majority of cases and they present a dramatic demonstration of the principle of segregation. Aberrant asci are produced, however, and their frequency varies with the gene concerned. In the case of $g$ (grey-spore), the most intensively studied locus, about 0·12% of the asci are abnormal (Table 6.1).

**Table 6.1**   Percentage frequency of abnormal segregation for the $g$ locus in *Sordaria fimicola*. The 4:4 segregations are abnormal with regard to spore arrangement and the frequency given is that for arrangements which could not be explained on the basis of nuclear passing. (Based on data of Kitani, Y. *et al.* (1962) *Am. J. Bot.*, **49**, 697–706.)

| Ascus type | 5:3 | 6:2 | 4:4 |
|---|---|---|---|
| Black:Grey | 0·052 | 0·047 | 0·008 |
| Grey:Black | 0·01 | 0·006 | |
| Total | 0·062 | 0·053 | 0·008 |

The most conspicuous aberrant asci are those in which 1:1 segregation does not occur. These generally show 5:3 or 6:2 segregation but 7:1 asci have been observed. Kitani, Olive and El-Ani, using suitably marked stocks, subjected asci showing aberrant segregation for the $g$ locus to tetrad analysis and found that:

1. The recombination values for short regions on either side of the $g$ locus were abnormally high when this locus showed abnormal segregation (negative interference). This effect was greater in 5:3 than 6:2 asci.
2. In 5:3 but not 6:2 asci, a lower than expected frequency of recombinants occurred in short, nearby but not adjacent regions e.g., $sp$-$mat$ and $mi$-$mat$ (positive interference).
3. Recombination in longer, more distant regions (e.g. $cor$-$st22$, centromere-$sp$ or -$mi$) conformed with expectation.
4. In almost all cases the chromatids which showed recombination in the immediate vicinity of the $g$ locus were those involved in the abnormality of segregation (see below), and
5. Although it is very close to $g$, the $mat$ locus behaved normally even when the segregation of $g$ was aberrant.

These observations clearly show that the events leading to abnormal segregation are highly localized and associated with unexpectedly high rates of recombination in the immediate vicinity. The extent of the elevated recombination can be seen from a comparison of the linkage values obtained from the selected aberrant asci, on the one hand, and those found when all asci are considered on the other (Fig. 6.1). The magnitude of the negative interference is dramatically illustrated also by the fact that double cross-overs in the *mi-cor* interval of aberrant asci were more than a hundred times more common than expected on the basis of the standard map. It will be appreciated that the very high re-combination frequencies in the vicinity of *g* in selected asci does little to distort the map based on random spores because the aberrant asci represent only 0·12% of all asci. In fact, cross-overs in the aberrant asci constitute less than 1% of the total cross-overs in the *g-cor* intercept for example.

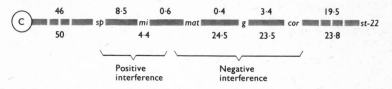

**Fig. 6.1** The *g* linkage group in *Sordaria fimicola*. The map distances given above the line are those obtained from unselected meiotic products, the values below the line are those for asci showing aberrant 5:3 segregation for the *g* locus. In these the data for *mi* and *mat* were not numerous and they have been pooled. (Based on data of Kitani, Y. *et al.* (1962), *Am. J. Bot.*, **49**, 697–706.)

Three other important observations were made. First, although aber-rant segregation is correlated with a high frequency of nearby recombina-tion, the latter is not a necessary pre-requisite for the former. Thus, 22 out of 74 asci showing 5:3 segregation did not show detectable crossing-over in either the *g-cor* or the *g-mi* or *-mat* intervals. Second, of those 17 asci in the same sample which had crossing over in both these intercepts, 14 showed 2-strand doubles and only 2 had 3-strand doubles, the aberrant chromatid being involved in both cross-overs in all 16. The remaining ascus showed 2-strand double crossing-over in which the aberrant chromatid was not involved in either exchange. Third, although 5:3 and 6:2 asci were equally frequent, the wild-type black spore was in the majority in both kinds of aberrant segregation (Table 6.1).

Asci with four spores of each type, though normal in respect of the segregation ratio, were sometimes abnormal in respect of spore arrange-ment. Some of these were due to spindle over-lap or nuclear migration, but in other cases the normal behaviour of both linked and unlinked markers showed that gross chromosome or nuclear abnormalities were not responsible for their production. Indeed, the nature of the sequences themselves frequently rendered unlikely an explanation based on nuclear passing. These aberrant 4:4 asci, like the 5:3 types, thus showed segregation at the third, presumably mitotic, division in the ascus (post-meiotic segregation). One case of 7:1 segregation was observed which also requires post-meiotic segregation. Finally, other spore colour mutants differ from *grey* not only in the over-all frequency of aberration but in the type of aberration observed. For example, the *mummy* mutation shows only 6:2 abnormalities and at a rate of 1 per 1500 asci.

## The *Ascobolus* affair

In *Ascobolus immersus* over 2000 mutations affecting the colour and distribution of the spore pigments have been isolated. Each of these generally gives a 1:1 segregation in crosses with wild-type and so single point mutations or deletions appear to be involved in every case. Crosses between the mutants themselves allow their classification into a number of series. Mutations in different series appear to involve different genes while those within a series are closely clustered on the linkage map. Consequently, most of the asci produced following crosses between intra-series mutants are expected to contain only mutant spores, two of one parental-type and two of the other. On occasion, however, exchange between mutant sites which are not strictly allelic and do not overlap is expected to give wild-type recombinants. In the event of normal, reciprocal recombination such asci containing wild-type spores should also contain doubly mutant spores as well as the mutant parental-types, all in equal numbers (Fig. 6.2). The various mutant forms would be phenotypically indistinguishable and the phenotypic ratio in the ascus would be 6:2 in favour of the mutant phenotype. But although they are indistinguishable in appearance, the single and double mutants are expected to behave differently in back-crosses with the two parents (Fig. 6.2).

Lissouba, Rizet and their co-workers have studied the behaviour of these spore-colour mutants in *Ascobolus* over a long period of time. The results obtained are extensive and the principal features of the analysis can be illustrated by reference to selected mutations in three different series.

SERIES 46    Fourteen pairwise crosses involving 6 mutants of this series were studied in detail and they all gave rise to some 6:2 asci. Detailed analysis of the mutant spores in these asci showed the following results in every case:

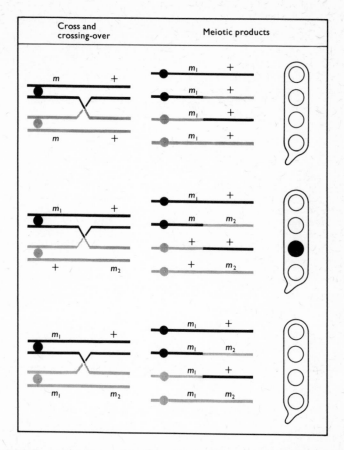

**Fig. 6.2** The expected consequences of reciprocal crossing-over between two marked sites within the same functional unit. (Top) Mating identical mutants gives homozygotes from which no detectable recombinants can arise (4:0 asci). (Centre) Crossing two phenotypically alike but not strictly allelic mutants gives double heterozygotes. Crossing-over between the two sites gives complementary wild-type and doubly mutant products (3:1 asci). (Bottom) Back-crossing double mutants to either parental type single mutant gives single heterozygotes in which crossing-over can reproduce only the immediate parents.

1. The mutant spores were always of one parental type or the other; no doubly mutant spores occurred. Thus, in no case could the production of the wild-type pair be attributed to reciprocal recombination (Fig. 6.3).

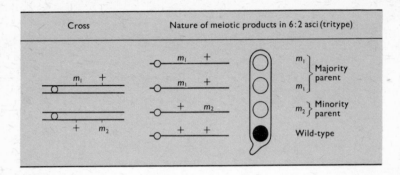

**Fig. 6.3** The nature of the segregants in the rare 6:2 asci formed following crossing between two spore colour mutants belonging to Series 46 in *Ascobolus immersus*. Note the absence of the doubly mutant product which is the expected complementary product of the wild-type following reciprocal exchange (cf. Fig. 6.2 centre). The parental type represented in the minority among the mutant spores in 6:2 asci corresponds with the locus showing abnormal 3:1 segregation.

2. The frequencies of 6:2 asci found in various crosses were roughly additive and a linear conversion map based on them could be constructed (Fig. 6.4).

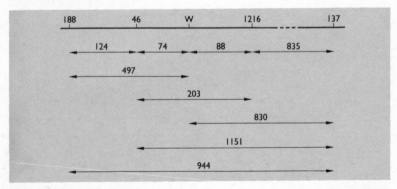

**Fig. 6.4** The distribution of the Series 46 spore colour mutant sites in *Ascobolus immersus* based on the frequency of 6:2 asci per 100,000 in mutant × mutant crosses. All these aberrant asci were the result of conversion. The values are approximately additive but note indication of map expansion.

3. In each and every cross the mutant site showing abnormal segregation in the double heterozygote was that to the 'right' on the conversion map. For example, in crosses with 188 or 46 the W mutation behaved abnormally but in association with 1216 and 137, the segregation of W

**Table 6.2** The nature of the abnormal 6:2 asci produced following crossing between mutants within Series 75 in *Ascobolus immersus* and the order of the mutant sites based on the frequency of these asci (crossing-over and conversion combined). (Data of Lissouba, P. *et al.* (1962) *Adv. Genet.,* **11**, 343–80.)

Series 75 mutant sequence:

231 |—10·4—| 322 |—14·2—| 278 |—16·4—| 147 |—20·4—| 1987

| Cross | No. of 6:2 Asci analysed | Cross-over asci (Double mutant present) | Conversion asci (Double mutant absent) Majority parent | |
|---|---|---|---|---|
| | | | Left marker | Right marker |
| 231 × 322 | 11 | 0 | 7 | 4 |
| 231 × 278 | 14 | 1 | 9 | 4 |
| 231 × 147 | 16 | 4 | 11 | 1 |
| 231 × 1987 | 17 | 5 | 11 | 1 |
| 322 × 278 | 16 | 3 | 7 | 6 |
| 278 × 147 | 19 | 3 | 12 | 4 |
| 278 × 1987 | 18 | 3 | 12 | 3 |
| 147 × 1987 | 17 | 1 | 8 | 8 |
| Total | 157 | 39 | 84 | 34 |

(84 and 34 combined: 118)

was normal. Thus, the mutant site sequence based on the incidence of 6:2 asci coincides with the series based on abnormal segregation. This order was confirmed in 48 asci from 11 crosses involving 6 other mutants in the series.

SERIES 75    The results obtained for this series were rather different and they exemplify the other extreme situation in regard to intra-series crosses. Thus:

1. While the majority of the 6:2 asci were the result of conversion about 30% of them contained doubly mutant spores and could, therefore, be attributed to reciprocal recombination.
2. On the basis of the frequency of 6:2 asci produced in pairwise crosses, whether by conversion or reciprocal exchange, the mutants could be arranged in sequence. This map, being based on the joint frequency of conversion and crossing-over, is complex and the values are only approximately additive. This is not surprising because the relative frequency of conversion to crossing-over varies widely between crosses (Table 6.2). However, support for the sequence comes from the fact that:
3. The mutant site showing abnormal 3:1 segregation in 6:2 asci was not constant for a given cross but in none of the crosses did the left hand marker behave abnormally more often than that on the right (Table 6.2).

The differences between the results for Series 46 and 75 probably depend, at least in part, on the greater length of the 75 region. This is suggested by the higher over-all frequency of 6:2 asci in crosses involving mutants of Series 75. What is more, this series has more known mutants. Support for this view comes also from the results obtained for Series 19.

SERIES 19    The results obtained for this series were intermediate between those obtained for Series 46 and 75. Thus, the mutants of Series 19 could be divided into groups. Crosses within groups gave results which closely approached those for Series 46. But inter-group crosses were more similar in their outcome to those in Series 75. The frequency of 6:2 asci found for Series 19 crosses, and the number of mutants known in the series, suggest that it is intermediate in length between 46 and 75.

Since wild-type spores can be produced in the absence of the expected doubly mutant reciprocal product, the reverse situation is expected to obtain also. However, asci of this type are expected to contain only spores with the mutant phenotype so their detection involves laborious analysis. In fact, the spores from 200 asci of the 8:0 type were analysed by back-crossing to the parent and no asci with doubly mutant spores were found. But such spores, produced by non-reciprocal conversion, were eventually found.

Further, monohybrid zygotes produced by wild-type/mutant crosses gave rise to some asci with 6 wild-type: 2 mutant and others with 2 wild-type: 6 mutant spores though not always in equal numbers. These are comparable with the crosses described above for *Sordaria*. Furthermore, the frequency of 6:2 asci found in these *Ascobolus* crosses increased as the location of the mutant site moved to the right of the conversion map.

Thus, these investigations suggest the existence of segments with the following properties:

1. Recombination within the segment is never reciprocal and always involves conversion.
2. Conversion frequencies for regions within the segment are additive.
3. Conversion frequencies for mutations within the segment increase progressively as the location of the site considered moves to the 'right' of the conversion map.

In acknowledgement of the polar properties of this level of genetic organization, Lissouba and Rizet suggested the term polaron for such a segment. On this basis, the mutations considered above in Series 46 belong to a single polaron, those within groups of Series 19 belong, for the most part, to the same polaron also but different groups represent different but linked polarons. To some extent, therefore, polarity extends beyond the limits of the polaron. This is seen also in Series 75 which covers more than one polaron. The polaron concept implies other features, for example, the restriction of reciprocal crossing-over to the junctions between polarons.

## The *Neurospora* story

By tetrad analysis, Case and Giles studied the results of a cross in *Neurospora* which combined certain features of the *Sordaria* and *Ascobolus* investigations. Thus, as in *Ascobolus*, asci produced by mutant × mutant crosses were examined with regard to the segregation of the mutant sites involved. These were very closely linked and within a gene (*pan-2*), the wild-type form of which confers pantothenate independence. Further, as in the *Sordaria* study, markers flanking the *pan* region were introduced into the cross.

Of the 867 asci subjected to tetrad analysis, 856 were ditype containing the two parental types of singly mutant *pan* spores in equal numbers. In other words, although the two mutant sites are not strictly allelic, they segregated in 98·7 per cent of the asci because no detectable recombination had occurred between them.

On a tetrad basis, eight of the remaining eleven asci each contained one pantothenate-independent spore indicative of such an exchange. However, only two of these asci contained also the expected product of reciprocal exchange, namely, the double mutant. In fact, these two asci were tetratype containing the two parental-type single mutants, and the two products of apparently reciprocal recombination in equal numbers.

The other six asci which contained a wild-type spore were tritype; the double mutant was not present but the two parental-type, single mutants were represented, albeit unequally. In other words, in these asci one or other of the *pan* loci showed 3:1 segregation in favour of wild-type, but polarity of the type described in *Ascobolus* was not evident. These tritype asci were heterogeneous in respect of outside-marker recombination and some polarity was evident in this direction. When the above experiment was taken in conjunction with the results of the reciprocal cross, about half the asci showing abnormal segregation for one or other of the *pan* mutations also showed recombination for the closely linked (c. 8 units) flanking markers, the segregation of which was always normal. Thus, as in *Sordaria*, aberrant segregation was highly localized and frequently, but not invariably, associated with recombination in the immediate vicinity of the aberration.

### The *Saccharomyces* study

It was pointed out earlier that the observed recombination value for two mutant sites often tends to be less than the map distance obtained by adding the recombination values for intervening intercepts. This disparity is intensified by negative interference and reduced by positive interference.

It was also pointed out that, for mapping purposes, 3-point experiments are generally preferable to a series of 2-point crosses. Now in a 3-point cross, the recombination value for the outside markers equals the sum of the values for the two intervening regions when interference is positive and complete. But where double crossing-over occurs, the recombination value for outside markers is *less* than the map distance derived by addition. Further in 3-point experiments, it is arithmetically impossible for outside marker recombination to *exceed* the calculated map distance because any strand which is recombinant for the outside markers must be recombinant for one or other of the intervening intercepts.

However, certain experiments concerning intragenic mutations in yeast have shown that the recombination value for two markers in a 2-point experiment may be greater than the map distance between them when this is determined in a 3-point experiment involving an intervening marker. Holliday who first drew attention to this phenomenon of map

expansion consequently argued that the mutational differences themselves impeded pairing. He further suggested that the intensity of this
effect increased not only with the number of mutational differences but
inversely with their distance apart.

The various peculiarities which are observed in studies of intra-genic
events have been illustrated above by reference to particular experiments
and particular organisms. It must be stressed, however, that the phenomena revealed in these experiments have also been found in others involving the same organism or a different one. For example, negative
interference has been described in yeasts and species of *Aspergillus*, while
partial polarity has been found in fungi other than *Ascobolus* and in a
bacterium, *Streptomyces coelicolor*.

In fact, although many peculiar features are apparent in the above investigations they are not at all exceptional. On the contrary they are
typical of those obtained whenever and wherever intra-genic events are
studied by tetrad analysis. Differences are found of course depending on
the region concerned and the particular mutations considered but these
variations are ones of degree rather than kind.

Directly or indirectly experiments on intra-genic events show many
common features. Thus, post-meiotic segregation is common to the
*Sordaria* and *Ascobolus* studies, negative interference is seen in *Sordaria*
and *Neurospora* while all three show departures from the Mendelian
expectations, and so on. This suggests that different aspects of the same
process are being observed whatever the precise nature of the study. And
on the basis of these investigations the following general conclusions can
be reached:

1. Intra-genic 'recombination' generally involves a non-reciprocal process (conversion).
2. This process leads to numerically unequal segregation in respect of
   which intra-genic polarity may or may not obtain. Asymmetry may
   also be evident in that 3:1 and 1:3 segregation, for example, may not
   be equally frequent.
3. Intra-genic events may be associated with numerically equal or unequal post-meiotic segregation. Numerically unequal post-meiotic
   segregation also may be asymmetrical.
4. Aberrant segregation is highly localized and it involves only very short
   segments.
5. Intra-genic events are generally associated with a higher than expected
   frequency of recombination for closely linked outside markers. This
   negative interference is highly localized also and polarity may be
   observed in respect of outside marker recombination.
6. The strands which are subjected to intra-genic modification participate

more often than expected in outside marker recombination as well.
7. Intra-genic recombination appears often to be impeded by heterozygosity, an effect which may be revealed in map expansion.

## MOLECULAR MODELS

The mechanisms responsible for genetic exchange have long been in
dispute. But, until the recent tetrad analyses of genetic fine structure, recombination between homologous segments could be satisfactorily discussed in terms of a strictly reciprocal exchange of corresponding
segments. In fact, recombination between loosely linked sites can still be
considered in these terms whatever the precise nature of the underlying
mechanism. Further, a quantitative reciprocity is still evident because
deficiencies and duplications are avoided.

When the events which occur at and around sites of actual exchange are
analysed, it becomes clear that the recombinational process is not a point
phenomenon, as the older theories of breakage and reunion assumed. It
is rather but one aspect and consequence of events which extend over a
segment of the genetic material. Further, although these segments are
short in relation to the total map length, they are quite extensive when
considered in molecular terms.

Recent molecular models of recombination suggest that, after chromosome pairing, short regions of hybrid DNA are formed in which one
polynucleotide column is of maternal origin while the other is derived
from the paternal side. Various versions based on hybrid DNA have been
proposed and the one offered by Holliday will be considered here.[52]

Basically the model suggests the following sequence of events:

1. Simultaneous breakage at strictly corresponding sites in two single
polynucleotide columns of the same polarity from non-sister chromatids.
2. Unwinding (strand separation) of the broken columns from their unbroken complementary half-helices. The extent of unwinding is
variable but it proceeds in one direction from the initial break-point.
3. Union of the broken ends in new combinations to give a half-chiasma.
This involves rewinding and presents some topological problems.

With the formation of a half-chiasma, the stage is set for a variety of
consequences whose probability of occurrence depends on the exact
structure of the half-chiasma (Fig. 6.5).

### Restitution and reciprocal exchange

Let us assume in the first instance that the original maternal and
paternal duplexes were identical in the region over which hybrid DNA

was subsequently formed. Under these circumstances, base pairing in
the hybrid region can occur strictly in accordance with the principle of
complementarity though some mechanical distortion can be expected
in the region of the half-chiasma itself. In other words, although the
DNA is hybrid in being partly of paternal and partly of maternal origin,

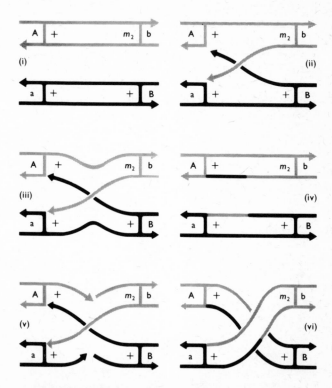

**Fig. 6.5** The Holliday Hypothesis of Crossing-over. Each strand represents a
single polynucleotide column of duplex DNA. Hydrogen cross-bonding is
not shown. A and B represent outside markers, + and $m_2$ sites within the
segment. Because of its distal location relative to the initial site of breakage, $m_2$
is less likely to be involved in the formation of hybrid DNA. (i–iii) Single-
strand breakage unwinding in one direction and cross-wise rejoining gives a
half-chiasma. (iv) The half-chiasma can be resolved by reunion following isolocus
breakage in the crossed-over strands at the point of the half-chiasma. This leads
effectively to the restoration of the original condition (but see Fig. 6.6). No
recombination of outside markers, normal segregation of sites within the segment.
(v–vi) The half-chiasma can be converted into a full chiasma by isolocus
breakage in the non-crossed-over strands. This leads to reciprocal recombination and normal segregation for markers within the segment and those flanking it.

it is not heteroduplex in structure. Paired chromosomes held together by a half-chiasma can separate only if the half-chiasma is resolved. This could be achieved in one of two ways (Fig. 6.5).

First, the polynucleotide columns which broke originally could break again but this time at the site of the half-chiasma. Subsequent unwinding and union in 'new' combinations would then restore the original condition and the whole sequence, in itself, would have no genetic consequences; the second series of breakage and union would cancel the first (but see Fig. 6.6).

Alternatively, however, the previously unbroken single polynucleotide columns could break at the half-chiasma and the ends thus produced could unite in new combinations. In this event the second series of breaks and unions would confirm the first and convert a half-chiasma into a complete one. The result would be normal segregation and reciprocal exchange (Fig. 6.5).

## Post-meiotic segregation

Let us now suppose that the hybrid DNA actually spans a site of difference between the maternal and paternal molecules. In this event the hybrid DNA would be heteroduplex at the locus of heterozygosity because non-complementary base pairs would be brought into apposition (Fig. 6.6).

If this condition persists, then, following the resolution of the half-chiasma (one way or the other) and the anaphase separations, two of the four meiotic products would receive heteroduplex molecules. These are not scheduled to replicate until the interphase which occurs between the end of meiosis and the first post-meiotic mitosis. During this replication cycle, the columns of each heteroduplex are expected to separate and template for complementary half-helices according to normal base-pairing principles. But since the two columns in each heteroduplex are not wholly complementary to each other, the sister duplexes produced by post-meiotic replication are not expected to be identical. In other words, post-meiotic segregation is the anticipated consequence. Notice, however, that if this sequence is followed in its entirety, numerical segregation is expected to be normal (4:4) (but see Fig. 6.6).

## Unequal segregation

In the tetrad analyses described earlier (see p. 124) post-meiotic segregation was apparent but unequal segregation without post-meiotic segregation did occur. This suggests that if heteroduplex DNA is produced following half-chiasma formation, then the distortion owing to

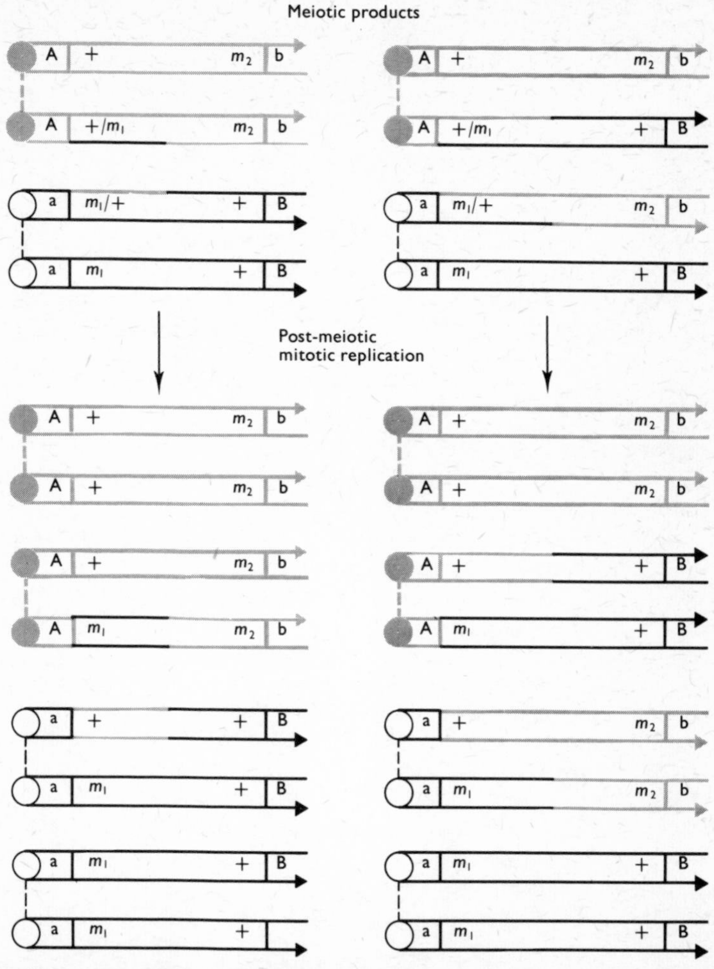

Meiotic products

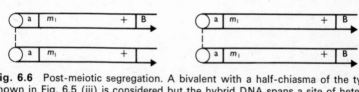

Post-meiotic
mitotic replication

**Fig. 6.6** Post-meiotic segregation. A bivalent with a half-chiasma of the type shown in Fig. 6.5 (iii) is considered but the hybrid DNA spans a site of heterozygosity and is, therefore, heteroduplex. All chromatids are shown. (Left) Consequences of resolving the half-chiasma by the method shown in Fig. 6.5 (iv). (Right) Consequences of confirming the half-chiasma as shown in Fig. 6.5 (v–vi). In both cases, post-meiotic segregation occurs if the heteroduplex persists. Both give recombination for inside markers but only a full chiasma recombines outside markers. Numerical segregation is normal (4:4) for all individual markers and hence phenotypic segregation is normal in wild-type × mutant crosses. But recombination between two markers within the segment gives reciprocal wild-type and doubly-mutant products, and hence, 7:1 phenotypic segregation in mutant$_1$ × mutant$_2$ crosses. Note that since $m_2$ lies further than $m_1$ from the initial break-point, the heteroduplex is less likely to include it.

the apposition of non-complementary bases can often be corrected before meiosis is completed. Correction may involve the excision of one or other base in the offending pair and its replacement by one which is complementary to the remaining member of the pair. However, excision may be followed by single-strand degradation for some distance on both sides of the distorted site. In this event some DNA synthesis would be required for the repair process.

If the two heteroduplex molecules are corrected to opposite parental types, segregation will appear to be normal. But if both heteroduplexes are converted to the same parental type a 3:1 (6:2) segregation should follow. Alternatively, if one hybrid molecule is corrected while the other remains heteroduplex a 5:3 ratio, which involves both unequal and post-meiotic segregation, should obtain (Fig. 6.7).

Now if the half-chiasma is the basis of both recombination and aberrant segregation, their frequent coincidence is explained and the fact that the same strands are generally involved in both is readily appreciated. But since the half-chiasma can be resolved without recombination and the heteroduplex can be corrected without aberrant segregation, an invariable association is not expected.

If it is further postulated that the probability and direction of correction depend on the precise constitution of the non-complementary base pairs in the heteroduplex DNA molecules, the different frequencies of unequal and post-meiotic segregation shown by different mutants can be accommodated. So too can inequalities in the frequency of 6:2 versus 2:6 and 5:3 versus 3:5 segregation (asymmetry) for a given mutation.

## Polarity

When other considerations are set aside, the probability that a given heterozygous locus will show abnormal segregation depends on the frequency with which it becomes involved in the formation of heteroduplex DNA. Further, if strand separation can cease at any point from the breakage site, the probability of involvement in heteroduplex DNA will depend on the distance between the locus and the breakage site. Furthermore, if strand separation proceeds in only one particular direction from the breakage site, it matters whether the locus lies to the 'left' or the 'right' of the breakage site.

However, if the frequency distribution of initial breakage is random along the length of the chromosome, so that breakage is no more likely to occur at one point than any other, then the distance between any given heterozygous locus and a site of actual breakage will vary continuously in the meiotic population. Consequently, the frequency of abnormal segregation shown by a particular locus should in no way be related to its location. But studies of *Ascobolus*, in particular, have shown that location is important in this connection. This difficulty can be overcome by assuming that there are preferred sites of breakage. These may not be numerous, they may be scattered at irregular intervals along the

| | | Correction prior to post-meiotic mitosis | | |
|---|---|---|---|---|
| Reciprocal heteroduplex molecules | | None — A‖C | To wild-type — A‖T | To mutant — G‖C |
| Correction prior to post-meiotic mitosis | None — G⃒T | A–T ● <br> A–T ● <br> A–T ● <br> G–C ○ <br> G–C ○ <br> A–T ● <br> G–C ○ <br> G–C ○ <br> 4:4 | A–T ● <br> A–T ● <br> A–T ● <br> A–T ● <br> G–C ○ <br> A–T ● <br> G–C ○ <br> G–C ○ <br> 5:3 | A–T ● <br> A–T ● <br> G–C ○ <br> G–C ○ <br> G–C ○ <br> A–T ● <br> G–C ○ <br> G–C ○ <br> 3:5 |
| | To wild-type — A⃒T | A–T ● <br> A–T ● <br> A–T ● <br> G–C ○ <br> A–T ● <br> A–T ● <br> G–C ○ <br> G–C ○ <br> 5:3 | A–T ● <br> A–T ● <br> A–T ● <br> A–T ● <br> A–T ● <br> A–T ● <br> G–C ○ <br> G–C ○ <br> 6:2 | A–T ● <br> A–T ● <br> G–C ○ <br> G–C ○ <br> A–T ● <br> A–T ● <br> G–C ○ <br> G–C ○ <br> 4:4 |
| | To mutant — G⃒C | A–T ● <br> A–T ● <br> A–T ● <br> G–C ○ <br> G–C ○ <br> G–C ○ <br> G–C ○ <br> G–C ○ <br> 3:5 | A–T ● <br> A–T ● <br> A–T ● <br> A–T ● <br> G–C ○ <br> G–C ○ <br> G–C ○ <br> G–C ○ <br> 4:4 | A–T ● <br> A–T ● <br> G–C ○ <br> G–C ○ <br> G–C ○ <br> G–C ○ <br> G–C ○ <br> G–C ○ <br> 2:6 |

**Fig. 6.7** The correction of illegitimate base pairs in heteroduplex DNA created by half-chiasma formation (see Fig. 6.6). A mutant (G-C) × wild-type (A-T) cross is considered. Only the two chromatids involved in the half chiasma are considered in detail but the contributions of the parental chromatids are included in the final ascus. Note that in the absence of other data, the post-meiotic segregations which give 4:4 segregation cannot be distinguished from normal pre- or post-reduction. 8:0 and 7:1 segregations could arise following four-strand double crossing-over. Asymmetry arises if the direction of correction is not random.

chromosome and although initial breakage is confined to such sites it can be expected to occur at only a fraction of them in any given cell. The notion of preferred breakage sites implies that there are structural discontinuities in the DNA molecule or else that certain base sequences can be recognized by a specific enzyme which effects breakage. These sites, whatever their nature, would coincide with the polaron junctions.

The *Ascobolus* data showed also that the frequency of 6:2 asci varied from cross to cross and, for mutations within a polaron, the production of wild-type spores depended entirely on gene conversion which involved a particular locus of a pair in dihybrid crosses. In other words, the frequency of abnormal segregation shown by a given mutant site depended on the second mutation which was combined with it in the dihybrid. Further, these frequencies were approximately additive so that the distance between the mutant sites was important—even though reciprocal recombination was not involved. The model outlined above has to be further elaborated to accommodate these findings.

The hybrid DNA model of recombination proposed by Whitehouse differs from that of Holliday in that the initial breakage is supposed to occur in strands of opposite polarity from non-sister chromatids.[94] Hybrid DNA is then produced by the annealing of the broken strands following unwinding. All existing models of recombination have unsatisfactory features but those based on hybrid DNA and half-chiasma formation provide the best basis for discussion and testing. At the genetic level the requirement is that a model provide a reasonable explanation of meiotic phenomena. But, clearly, even if this condition is fulfilled a model cannot be regarded as satisfactory until the underlying processes it invokes can be shown to exist.

In this connection the above model proposes that:

1. A specific breakage enzyme exist which can recognize particular sites, the so-called polaron junctions or recombinator regions, which are consequently subject to preferential breakage.
2. Strand separation is followed by the formation of hybrid DNA and half-chiasmata.
3. A repair system exists which can recognize and correct departures from the regular double DNA helix.
4. Repair may involve more or less extensive DNA synthesis which takes place after the cells have entered division.

We cannot enter into the evidence for these processes here but let it be said that their existence can be supported.[28]

*Interlude*

## THE GENE CONCEPT

The zygotes, merozygotes or mixed infections produced on combining certain mutants and their wild-type progenitors give rise to only parental-type (wild or mutant) segregants even when large numbers of offspring are examined. Such a result indicates that the genetic difference between the mutant concerned and its wild-type progenitor is indivisible by recombination. In other words, the mutated site can be regarded as a point so far as its behaviour in heredity is concerned.

Of course, in physical terms the region of difference on the genetic material must have a finite length. In chemical terms, the mutational change could conceivably involve one or more base locations for, clearly, a segment will appear as a point on a linkage map if recombination cannot occur within it. Similarly, unless two sites or regions are separable by recombination they cannot be distinguished as such in breeding experiments.

Conversely, the existence of a genetic site becomes apparent only if it mutates, and distinguishing between two sites requires not only recombinability but independent mutability.

In genetic analysis, then, the unit of mutation and the unit of transmission (or recombination) cannot be considered separately because the one is relevant to the detection of the other. It will be appreciated also that a presence/absence difference will behave in a unitary way in heredity whatever the extent of the absence for, clearly, a deficiency cannot be divided by reciprocal exchange! Similar considerations apply to various other gross structural rearrangements.

However, we have seen that independently-mutable, apparently-indivisible sites separable by the breakage involved in crossing-over, chromosome rearrangement or conjugal transfer, can be mapped relative to one another. Under normal circumstances, essentially linear, approximately additive maps are obtained when these mutant sites are plotted at distances apart proportional to the non-random frequency with which their linkage is broken. Further, the site-sequence in these maps corresponds with that on the physical gene string.

Maps of this type illustrate one aspect of genetic analysis, namely, the determination of the hereditary relationship between identifiable loci. Sites which are subject to alternative inheritance, and are therefore genetically allelic, map at the same locus even though they may not occupy exactly corresponding homologous positions on the genetic material. Sites which are not genetically allelic, on the other hand, are mapped at different loci and these, of course, must be structurally non-allelic also.

The other aspect of genetic analysis is concerned with defining the functional relationship between genetically identifiable loci.[53] Does independence in mutation and separability in heredity indicate an autonomy of function?

As early as 1909, Archibald Garrod, an English knight and physician, recognized that many congenital diseases in man were associated with abnormal enzyme activity.[36] Some years later, studies by Onslow and Bassett on the anthocyanin pigments of plants showed that corolla colour varied with the ability to carry out certain specific and identifiable chemical reactions and that this was genetically controlled in *Antirrhinum*. But the major break-through in the elucidation of gene action began about 30 years ago with the work of Beadle and Tatum who formulated the one-gene/one-enzyme hypothesis.[8] More recent work has extended knowledge of protein structure and function and clarified the relationship between genes and both enzymic and non-enzymic proteins.

# PART III

## THE FUNCTIONAL ORGANIZATION OF THE GENOTYPE

# 7

# The Expression of Genetic Information

## THE STRUCTURE OF PROTEINS

Proteins, like nucleic acids, are polymers and they consist of a sequence of conjoined α-amino acid (AA) units of the general type $R.CH.NH_2.$ COOH. Only about twenty amino acids take part in protein production although more are known to occur even in biological systems (Table 7.1).

α-AA units are capable of forming polypeptide chains through the association of the amino ($NH_2$) group of one AA and the hydroxyl (OH) group of the carboxyl (COOH) radical of another (Fig. 7.1), and so on. Each chain can be considered as starting from an AA with a free-$NH_2$ group (N-terminal) and ending with one whose carboxyl (COOH) group remains free (C-terminal). But even these terminal groups do not necessarily remain unrecombined, for they may form linkages with molecules other than AA.

In the formation of polypeptide chains each AA contributes an identical peptide group to the backbone of the molecule (—NH.CH.CO—) together with a distinguishing (R) group. Each kind of protein thus has a unique number and sequence of such R groups which, in the last analysis, determine the functional possibilities of the molecule. Actually, one of the 20 AA, proline, has an imino (NH) rather than an amino ($NH_2$) group. At sites where proline is inserted into a polypeptide chain, then the —NH—CH(R)—CO— sequence becomes $>$N—CH(R)—CO—,

**Table 7.1**   The amino acids of proteins

| Amino acid | Formula | Residue symbol | Mol. wt. |
|---|---|---|---|
| **The neutral amino acids** | | | |
| a. Glycine | $H_2N \cdot CH_2COOH$ | -gly- | 75 |
| b. Alanine | $H_2N \cdot CH \cdot COOH$<br>      &#124;<br>     $CH_3$ | -ala- | 89 |
| c. Valine | $H_2N \cdot CH \cdot COOH$<br>      &#124;<br>    $CH(CH_3)_2$ | -val- | 117 |
| d. Leucine | $H_2N \cdot CH \cdot COOH$<br>      &#124;<br>   $CH_2 \cdot CH(CH_3)_2$ | -leu- | 131 |
| e. Isoleucine | $H_2N \cdot CH \cdot COOH$<br>      &#124;<br>$CH_3 \cdot CH \cdot C_2H_5$ | -ileu-<br>(-ile-) | 131 |
| f. Phenylalanine | $H_2N \cdot CH \cdot COOH$<br>    $CH_2$ ⬡ | -phe- | 165 |
| g. Proline | $H_2C \longrightarrow CH_3$<br>$H_2C \diagdown_N{}^{\diagup} CH \cdot COOH$<br>     H | -pro- | 115 |
| h. Tryptophan | ⬡$\diagup C \cdot CH_2 \cdot CH \cdot COOH$<br>      CH    $NH_2$<br>      N<br>      H | -try-<br>(-trp-) | 204 |
| i. Serine | $H_2N \cdot CH \cdot COOH$<br>      &#124;<br>   $CH_2OH$ | -ser- | 105 |
| j. Threonine | $H_2N \cdot CH \cdot COOH$<br>      &#124;<br>  $CHOH \cdot CH_3$ | -thr- | 119 |
| k. Methionine | $H_2N \cdot CH \cdot COOH$<br>      &#124;<br> $(CH_2)_2 \cdot SCH_3$ | -met- | 149 |
| l. Cystine | $COOH \qquad COOH$<br>    &#124;        &#124;<br>$H_2N \cdot CH \qquad CH \cdot NH_2$<br>    &#124;        &#124;<br>$CH_2 \cdot S \cdot S \cdot CH_2$ | -cys<br> &#124;<br>-cys | 240 |

**Table 7.1** The amino acids of proteins (continued)

| Amino acid | Formula | Residue symbol | Mol. wt. |
|---|---|---|---|
| m. Asparagine | $H_2N.CH.COOH$ <br> $\quad\vert$ <br> $CH_2.CONH_2$ | -aspN- <br> (-asn-) | 132 |
| n. Glutamine | $H_2N.CH.COOH$ <br> $\quad\vert$ <br> $CH_2.CH_2.CONH_2$ | -gluN- <br> (-gln-) | 146 |
| **The acidic amino acids** | | | |
| o. Aspartic acid | $H_2N.CH.COOH$ <br> $\quad\vert$ <br> $CH_2COO^-$ | -asp- | 133 |
| p. Glutamic acid | $H_2N.CH.COOH$ <br> $\quad\vert$ <br> $CH_2.CH_2.COO^-$ | -glu- | 147 |
| q. Tyrosine | $H_2N.CH.COOH$ <br> $\quad\vert$ <br> $CH_2$—⬡—$O^-$ | -tyr- | 181 |
| r. Cysteine | $H_2N.CH.COOH$ <br> $\quad\vert$ <br> $CH_2.S^-$ | -cys- | 121 |
| **The basic amino acids** <br> s. Histidine | $H_2N.CH.COOH$ <br> $\quad\vert$ <br> $CH_2$—$C$═$CH$ <br> $\quad HN\quad NH$ <br> $\quad\quad\searrow C \swarrow^+$ <br> $\quad\quad\quad H$ | -his- | 154 |
| t. Lysine | $H_2N.CH.COOH$ <br> $\quad\vert$ <br> $(CH_2)_4.NH_3^+$ | -lys- | 146 |
| u. Arginine | $H_2N.CH.COOH$ <br> $\quad\vert$ <br> $(CH_2)_3.NH.C$═$NH_2^+$ <br> $\quad\quad\quad\quad\vert$ <br> $\quad\quad\quad\quad NH_2$ | -arg- | 174 |

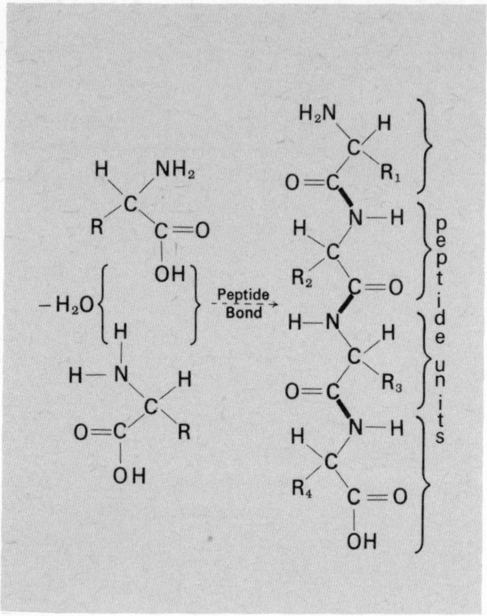

**Fig. 7.1** The basic formula of amino acids and the primary structure of poly-peptide chains. Peptide bonds shown in bold.

this grouping itself forming part of a pyrrolidine ring. One of the consequences of this is a change in the direction of the main chain.

The linear sequence of the AA components constitutes the primary structure of a polypeptide. The functional groups of the polypeptide chain have a tendency to form H-bonds, thus changing their own geometrical arrangements and so indirectly modifying the chemical and biological activities of the chain. For example an H-bond will form between a —CO group and an —NH group when these face one another and the distance between them is about 2·8 Å while for an O—H····O bond the minimum distance required is about 2·6 Å (Fig. 7.2). Secondary protein structure is thus largely governed by primary structure.

H-bonds such as these may form between groups on different polypeptide chains giving rise to so-called β-configurations. These are fully extended chains joined by N—H····O=C bonds into sheet-like structures. β-arrangements are favoured by the presence of large numbers of glycine and alanine residues and, in nature, occur chiefly in silk proteins.

The formation of H-bonds between groups in the same chain, however, twists the chain into an α-helix (Fig. 7.2). This is the most important regular arrangement of the polypeptide chain, though hardly any proteins exist purely in this form. The AA proline, for example does not fit into the α-helix pattern because the nitrogen group of proline is unable to form the crucial H-bond. In haemoglobin, therefore, proline groups occur at points in the molecule where the chain changes direction and hence abandons the α-helix form.

In many proteins a higher order tertiary structure accounts for the

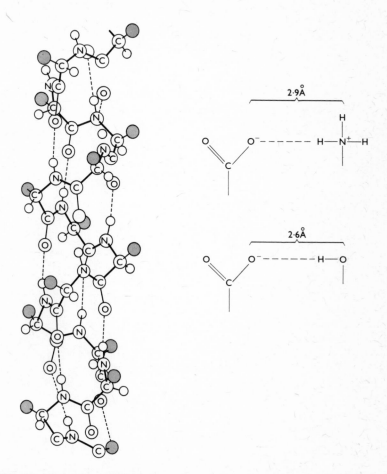

**Fig. 7.2** The alpha helix and the nature of the hydrogen bonds which contribute to secondary and tertiary protein structure. R groups shown as solid circles, H as small open circles. Main backbone indicated by solid bonds, hydrogen bonds by broken lines.

degree of compactness they adopt in aqueous solution. This tertiary structure depends on the manner in which the coiled polypeptide chain is packed within the molecule. The linkages responsible for the stability of tertiary structure are intra-chain bonds, especially disulphide bridges which serve to close loops in a polypeptide chain. The formation of these is dependent on the presence in cysteine residues of the very reactive sulphydryl (thiol) group which readily forms a disulphide (—S—S—) bridge on oxidation. Thus, when they confront each other, two distinct cysteine residues from the same polypeptide chain, or even different chains, may become linked to form the double AA cystine. For example, insulin, the first protein whose primary structure was determined, contains two polypeptide chains held together by two —S—S— bridges. The A chain has 21 AA residues and the B chain 30 (Fig. 7.3). Ribonuclease, on the other hand, contains four intra-chain disulphide bridges which play an important role in determining the three-dimensional arrangement of the molecule.

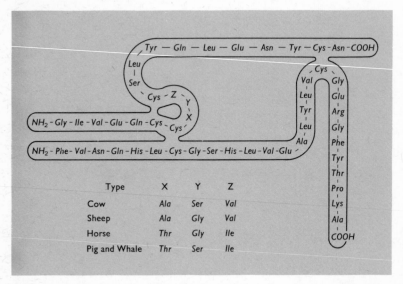

| Type | X | Y | Z |
|------|-----|-----|-----|
| Cow | Ala | Ser | Val |
| Sheep | Ala | Gly | Val |
| Horse | Thr | Gly | Ile |
| Pig and Whale | Thr | Ser | Ile |

**Fig. 7.3** The primary structure and inter-specific variation of the polypeptide chains in heteromultimeric insulin. The folding indicated is arbitrary.

The formation of a tertiary structure imposes new restrictions on the chemical reactivity of the basic polypeptide chain, adding still further to the limitations created by secondary folding into α-helices. But the masking of so many potentially reactive sites in no sense renders the protein molecule chemically inert since numerous groups may still remain available for combination. Tertiary structure permits interaction over a vast surface and creates interstices between the polypeptide chains into which particular reaction patterns can fit with optimal accuracy. Thus, the geometry of the gaps in the system may be as important functionally as that of the polypeptide framework itself.

The cross-linkages responsible for secondary and tertiary protein structure are largely intra-chain bridges. Equivalent inter-chain bridges may link together two or more otherwise independent polypeptide chains and so join sub-units into a complex quaternary or multimeric edifice. Most proteins with a molecular weight higher than 20,000 possess such a quaternary structure and are composed of more than one polypeptide chain (Table 7.2).

**Table 7.2** Composition and size of some well-known proteins. (After Schultze, H. E. and Heremans, J. F. (1966) *Molecular Biology of Human Proteins*, Vol. 1, Elsevier Publishing Co., London.)

| | Protein | Source | No. of poly-peptide chains | Mol. Wt. of chains | No. of AA |
|---|---|---|---|---|---|
| **1. Eucaryotes** | 1. Cytochrome-c | Human heart | 1 | 12100 | 104 |
| | 2. Ribonuclease | Bovine pancreas | 1 | 15800 | 124 |
| | 3. Myoglobin | Sperm whale | 1 | 17816 | 153 |
| | 4. Chymotryp-sinogen | Bovine pancreas | 1 | 2500 | 246 |
| | 5. Haemoglobin | Human blood | $2\alpha$ $2\beta$ | c.1600 | 141 146 |
| | 6. Lactate dehydrogenase | Human blood | 4 | 3400 | — |
| **2. Procaryotes** | 7. TMV | | 2300 | 17420 | 157 |
| | 8. Turnip yellow mosaic virus | | 120 or 180 | 21300 | 200 |
| | 9. Bacteriophage T2r | | 2000 | 80000 | — |

The haemoglobin (Hb) molecule, which illustrates some of the characteristic features of quaternary structure, consists of 4 polypeptide chains, two of one kind and two of another, each with a molecular weight of about 16,000 (total M. Wt. 64,500), and each associated with one haem group. The four chains are disposed in a tetrahedral arrangement but barely touch one another, being held together mainly by H-bonds and, to some extent, by electrostatic attraction.

Two other interesting and important properties of proteins merit attention:

1. The biological activity of a given protein usually resides in a particular region of the polypeptide chain—the so-called active site. The active sites of different proteins with comparable biological activities tend to show related, or even identical, AA sequences. Chymotrypsin, for example, has the enzymatic function of hydrolysing C—X bonds of the type

$$\overset{\displaystyle R}{\underset{\displaystyle |}{}}$$
$$-CO-NH-CH-CO-X$$

where R is a benzyl or a p-hydroxybenzyl residue and X may, but need not be, the AA group of a polypeptide. Tests with enzyme inhibitors, with agents modifying AA and with partial digestion, have shown that this enzyme has two kinds of active site, one for recognition of the substrate, the other for hydrolysis. Where hydrolysis is concerned, only two of the 242 AA in the molecule, one of the 30 serines and one of the 2 histidines, seem to be critically involved. The two crucial AA occur in distinct polypeptide chains. Protein molecules are thus highly differentiated functionally, each function involving some part of the molecule specifically. It must be emphasized that the basis of the functional organization of a protein is not wholly understood but it is clear that only a small part of the total AA complement is critically involved in function.

2. The species-specificity of proteins also appears to be associated with well-defined sequences of AA as is well demonstrated by many enzymes and hormones. Moreover, the species-specific sequences appear to be distinct from those which determine reaction specificity. For example, species-specific AA sequences are also the carriers of antigenic properties. Consequently most of the species-specificity of immunoglobins is located in peptide fragments that do not participate in antibody activity. For this reason it is possible with the aid of proteolytic enzymes to carve an immunologically active piece out of antibody molecules from the horse and to employ this fragment in human therapy with little of the adverse reactions usually encountered with species-foreign materials.

A considerable number of molecular variants in protein structure is now known. The commonest of these are allomers, that is molecular variants differing in some inessential detail(s) of their composition. For example, there are 104 AA in the polypeptide chain of most kinds of cytochrome-c though the chain is extended by 4-6AA at the N-terminal end in the case of yeasts and *Neurospora*. Certain sites, such as that of residue 89, exhibit many different AA changes, presumably because these changes do not disturb the secondary or tertiary structure of the protein to an extent that interferes with the specificity of its biological activity. Likewise 58 of the 158 AA sites in the coat protein of the *vulgare* and *dahlemense* strains of the RNA-containing TMV virus are subject to comparable changes (see also Fig. 7.3).

Allomerism is not confined to primary protein structure; it can extend even to the quaternary level. Quaternary allomers are proteins which differ by one or more molecular sub-units. Except in the case of the lamprey, the haemoglobin of which is monomeric (153 AA, M. Wt. 1700),

**Table 7.3** Primary allomerism in two tetrameric proteins.

| Allomer | | Polypeptide composition |
|---|---|---|
| A. Haemoglobin | 1. HbA | $\alpha_2^A \beta_2^A$ |
| | 2. HbA$_2$ | $\alpha_2^A \delta_2^A$ |
| | 3. HbF | $\alpha_2^A \gamma_2^F$ |
| | 4. HbH | $\beta_4^A$ |
| | 5. Hb-Barts | $\gamma_4^F$ |
| | 6. Hb-Portland | $\varepsilon_2^F \gamma_2^F$ |
| B. Lactate dehydrogenase | 1. LDH 1 | B$_4$ |
| | 2. LDH 2 | A$_1$ B$_3$ |
| | 3. LDH 3 | A$_2$ B$_2$ |
| | 4. LDH 4 | A$_3$ B$_1$ |
| | 5. LDH 5 | A$_4$ |

the haemoglobins of vertebrates are tetramers consisting of four globin chains. These tetramers can be chemically dissociated into two dimers and thence into four monomeric units. In the most common form of human Hb, adult HbA, there are two α-chains (each of 141 AA) and two β-chains (each of 146 AA) so that the molecule may be written as $\alpha_2^A \beta_2^A$ (574 AA, M. Wt. 67,000). In the α and β polypeptide chains 68 AA are in identical positions and their secondary and tertiary structures are also very nearly the same. Normal adult human blood also contains HbA$_2$ ($\alpha_2^A \delta_2^A$) combined with HbA in the proportion of 1:25. The α and δ chains are closely related, differing only in respect of 8 or 10 AA. Yet a further allomer, HbF ($\alpha_2^A \gamma_2^F$) is found in foetal blood and this variant is replaced by HbA in the first six months of neonatal life.

Five variants are known also in lactate dehydrogenase (LDH) where they have been given the name of isoenzymes (=isozymes). LDH also is tetrameric, having a molecular weight of 34,000. Two sorts of subunits, A and B, have been recognized by their differing electrophoretic mobilities and the 5 allomers represent the 5 tetramers that can be assembled by combining the A and B sub-units in all possible numerical combinations (Table 7.3).

## GENES AND ENZYMES

The pathways which connect the primary products of the genetic material and the final phenotype are often long and tortuous. Their complexity increases with that of the organism. It is not surprising, therefore, that the major advances in the elucidation of this epigenetic edifice were made on micro-organisms.

The mould, *Neurospora crassa*, can live on a simple artificial medium containing inorganic salts, a source of organic carbon and the vitamin biotin. Evidently, it has the inborn capacity to synthesize all the other vitamins, amino acids and nitrogenous bases essential to normal development. On the assumption that this capacity was genetically conferred, Beadle and Tatum argued that mutation could lead to nutritional requirements over and above those satisfied by a minimal medium and that the nature of the biosynthetic block could be determined to some extent from the nature of the growth supplement which restored normal development.[8]

In fact, they found that irradiation increased the frequency of auxotrophic mutant strains which could not survive on minimal medium but developed normally when one or other supplement was added to it. The main conclusions reached in such studies as these can be illustrated by reference to the arginine-requiring mutants which have attracted the attention of many investigators (Table 7.4):

1. The metabolic blocks of hundreds of independently-isolated point mutants can be circumvented by an exogenous supply of arginine.
2. The mutants concerned can be classified genetically into a number of groups according to the location of the mutant sites. Ten distinct mutant clusters, shared by five of the seven chromosomes in the haploid complement, have been mapped so far and two others are as yet unplaced. This genetic classification is supported by complementation tests (see p. 181).
3. The arginine auxotrophs can be classified on functional grounds also. Thus, although they all grow when arginine is added, some will survive when citrulline is the only supplement, while added ornithine

**Table 7.4** The arginine pathway in *Neurospora*.

| Wild-type pathway | Mutant nos. | Requirement | Accumulated metabolite | Enzyme impaired or absent |
|---|---|---|---|---|
| ↓<br>ORNITHINE | arg 4-9 | Ornithine, Citrulline, or Arginine | — | — |
| Carbamyl Phosphate ⟍ ↓<br>CITRULLINE | arg-12 | Citrulline, or Arginine | — | Ornithine transcarbamylase |
| Aspartic Acid ⟍ ↓<br>ARGININO-SUCCINIC ACID | arg-1 | Arginine only | Citrulline* | Arginino-succinate synthetase |
| Fumaric Acid ← ⟍ ↓<br>ARGININE | arg-10 | Arginine only | Arginino succinic acid | Arginino-succinase |

* Slight accumulation of added Citrulline.

is adequate support for others. In each case, however, a single supplement is sufficient.

4. A further functional classification is possible in some cases. For example, while some of the mutants which survive only on arginine accumulate argininosuccinic acid, others do not.

5. In certain cases, enzymes which are believed to be involved in arginine synthesis and which can be shown to be active in cell-free preparations of wild-type strains, prove to be absent, inactive or less active in auxotrophic mutants. Further, each point mutant is generally defective only in respect of one enzyme and the same enzyme is impaired in independently-isolated mutants belonging to the same group.

Studies of this kind can help to elucidate metabolic pathways but they have also provided the basis for a general hypothesis of gene action which holds that:

1. Genes can influence the peripheral phenotype by affecting the structure, function or production of enzymes.

2. The genetic material is unifunctional in these respects, a single point mutation affecting only a single enzyme and, hence, a single step in a biochemical pathway (but see p. 188). Consequently, the supply of a single substance, the end product or an intermediate beyond the block, can potentially alleviate the impediment. Of course, in theory, a supply of enzyme would also be sufficient medication.

This one-gene/one-enzyme hypothesis was subsequently extended to include proteins in general. However, experimental results are not always as clear-cut as those shown in Table 7.4 which have been selected and simplified for clarity. The following examples, again from *Neurospora*, illustrate some of the complications.

*Multiple requirements*

1. Common precursor: Wild-type strains can effect the reduction of dehydroshikimic acid to shikimic acid but reductase activity cannot be detected in the so-called *arom-1* mutants. These auxotrophs will not grow unless they are supplied with tryptophan, tyrosine, phenyl-alanine *and* p-aminobenzoic acid. However, they do not violate the unifunctional aspect of the above hypothesis because the four aromatic compounds they require are all produced from a common precursor, shikimic acid. Thus, single point mutations having a single primary effect on reductase activity confer secondary requirements for a range of substances.

2. Common enzymes: The synthetic sequence leading from aceto-lactic acid to valine and that from α-acetyl α-hydroxy butyric acid to isoleucine involve parallel steps. In fact, only a single additional methyl group distinguishes the precursor, intermediates and end product of the second sequence from their counterparts in the first. Now *iv* mutants have a requirement for both valine and isoleucine.

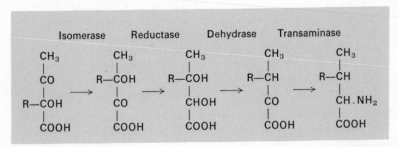

**Fig. 7.4** The parallel steps in the final stages of valine and isoleucine synthesis. In the former the R radical is represented by a methyl group ($-CH_3$), in the latter by an ethyl ($CH_2-CH_3$) group.

But this double need depends on a single primary effect on one or other of the enzymes which are common to both sequences (Fig. 7.4).

## Multiple enzymes

1. Common component: The synthesis of leucine involves amongst other things the conversion of α-isopropylmalate into its β-isomer and the dehydrogenation of the latter to give α-ketoisocaproate, the keto-acid analogue of leucine. Certain, single mutations (*leu-3*) on chromosome I affect both these steps although they are catalysed by different enzymes. It has been suggested, however, that both the enzymes concerned are composed of more than one kind of polypeptide chain (heteromultimers, see p. 180) and that the chain specified by the *leu-3* region is common to the two enzymes. A unifunctionality of primary action can be maintained on this basis.

2. Differential Component: If the above explanation of the *leu-3* mutants is correct, it follows that the polypeptides peculiar to the isomerase or the dehydrogenase must be specified by regions which are distinct both from each other and from *leu-3*. In fact, *leu-2* mutations (Chromosome IV) and *leu-1* mutations (Chromosome III) affect only the isomerase or the dehydrogenase respectively. Thus, where the enzymatically active protein is a heteromultimer, mutations at very different and even unlinked loci can affect a single enzyme and, hence, a single step in a biochemical pathway.

The above explanation of the *leu* mutants in *Neurospora* has not been established with certainty but the phenomena invoked have been clearly demonstrated in the case of haemoglobin (see p. 151). Further, many heteromultimeric enzymes are now known, but it appears that the genes which determine the structure of their component polypeptides are generally linked in tandem (see p. 180).

The exceptional results described on page 155 do not violate the one-gene/one-enzyme hypothesis although they appeared to do so at first sight. The above 2 exceptions, on the other hand, clearly do offend it but unifunctionality can still be maintained if the original proposal is modified to one-gene/one-polypeptide.

However, it is now quite clear that even the modified form of the hypothesis cannot accommodate certain classes of mutants. But these are special cases in that:

Either 1. The genes involved serve special functions in relation to gene transcription and translation (see p. 159),

Or 2. The mutational changes themselves are of a peculiar kind in relation to the sequential nature of polypeptide synthesis and the comma-free nature of the genetic code (see p. 188).

If we set aside these special cases for the present, there is ample evidence that so-called structural genes specify the primary structure of poly-peptides and that they are unifunctional in this respect.

The first experimental demonstration that genes control AA sequences in polypeptide chains came from the study of primary allomers in the haemoglobin (Hb) molecule of man. Here the specific primary structure of the α- and β-chains is controlled by two distinct and unlinked genes single mutations of which affect either the α- or the β-chain but not both. Thus:

$$\text{HbA} = \alpha_2^A\beta_2^A \begin{array}{c} \longrightarrow \text{HbS} = \alpha_2^A\beta_2^S \\ \longrightarrow \text{HbI} = \alpha_2^I\beta_2^A \end{array}$$

Some thirty-three variants of HbA have been chemically characterized and every case differs from the wild-type by the substitution of a single

**Table 7.5** Naturally-occurring mutational changes in the α-chain of human haemoglobin. Comparable mutants are known for the β- (38), γ- (3) and δ- (3) chains.

| AA site | AA change from | AA change to | Hb-type |
|---------|------|-----|---------|
| 1.  α-5 | ala | asp | J Toronto |
| 2.  α-15 | gly | asp | J Oxford, I Interlaken |
| 3.  α-16 | lys | glu | I I Texas |
| 4.  α-22 | gly | asp | J, Medellin |
| 5.  α-23 | glu | val | G Audhali |
| 6.  α-30 | glu | gln | G Honolulu, G Singapore, G Hong Kong |
| 7.⎫ α-47 | asp | gly | Umi, Kokura |
| 8.⎭ | asp | his | Sealy, Hasharan |
| 9.  α-51 | gly | arg | Russ |
| 10.⎫ α-54 | gln | arg | Shunonosecki |
| 11.⎭ | gln | glu | Mexico |
| 12.  α-57 | gly | asp | Norfolk, G Ibadan |
| 13.  α-58 | his | tyr | M Boston, M Osaka |
| 14.  α-68 | asn | lys | G Philadelphia, G Bristol, D alpha St. Louis, X, Stanleyville 1 and 2 |
| 15.  α-84 | ser | arg | E Tobicoke |
| 16.  α-87 | his | tyr | M Kankakee, M Shibata, M Iwate |
| 17.⎫ α-92 | arg | leu | Chesapeake |
| 18.⎭ | arg | gln | J Cape Town |
| 19.  α-115 | ala | asp | J Tongariki |
| 20.  α-116 | glu | lys | O Indonesia |

AA (Table 7.5). The β-chains, unlike the α-, are not produced during pre-natal life or early infancy when γ-Hb is in use. Consequently, drastic changes in the α-chain, unlike comparable changes in the β-chain, are likely to cause death *in utero*. This probably accounts for the fact that the known α-chain variations are less numerous than their β-chain counterparts. This example shows clearly that, even when different polypeptide chains are incorporated into one protein molecule, each chain is specified separately. It thus provides a dramatic demonstration of the one-gene/one-polypeptide hypothesis.[56]

The gene/polypeptide relationship has been defined even more precisely as a result of studies on micro-organisms which are more amenable to genetic analysis than man.[78] Perhaps the most significant investigations in the present connection are those on tryptophan synthetase in *E. coli*.[96] This enzyme is a heteromultimer consisting of two proteins, A and B, which are easily separable and recombinable *in vitro*. The A protein consists of a single polypeptide chain and the B component may be monomeric also. Mutants of the *tryp-1* group are defective in respect of the A chain while the B component is abnormal in mutants of the *tryp-2* group. *Tryp-1* and *tryp-2* mutant sites form two distinct but adjacent clusters on the linkage map. Consequently, deletions which span the junction between them affect both components of the enzyme (see p. 180).

The A polypeptide has a molecular weight of about 29,500 and consists of 282 amino acid residues, the sequence of which is mostly known. The amino acid sequence of about a dozen *tryp-1* mutants has been analysed also. In each case the defective A protein differs from the wild-type

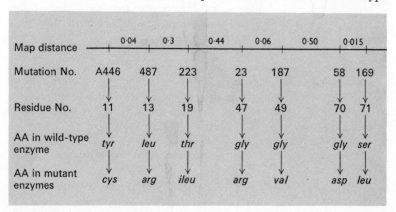

**Fig. 7.5** Colinearity of a gene and its polypeptide product in a 75 AA residue segment of the A-protein of tryptophan synthetase from *E. coli*. Other mutations affecting residues 47 and 70 are shown in Fig. 7.10. (Data of Yanofsky, C. *et al.*, (1964), *Proc. natn. Acad. Sci. U.S.A.*, **51**, 266–72.)

chain in respect of a single amino acid substitution.[97] But these studies extend those on haemoglobin in showing that:

1. There is a correspondence (*colinearity*) between the sequence of the mutatable sites on the linkage map and that of the amino acid substitutions they determine in the polypeptide chain (Fig. 7.5).
2. The map distances between the mutant sites are in good agreement with the spacing of the corresponding sites of amino acid substitution.
3. A given mutable site never affects more than one amino acid location, (but see p. 168).
4. The same amino acid location can be affected by distinct, though very closely linked, mutable sites (see p. 169).

## THE SYNTHESIS OF PROTEINS

Protein synthesis[5] involves the participation of RNA rather than DNA molecules and it takes place principally in the cytoplasm of cellular organisms. It will be recalled (p. 33) that three genera of non-genetic RNA can be distinguished on the basis of size, structure and function. All three are concerned with protein synthesis which is believed to involve the following stages in *E. coli*:

1. The binding of individual amino acids and ATP. This step is mediated by AA-specific activating enzymes (amino acyl RNA synthetases) and it results in the formation of aminoacyl-AMP-enzyme complexes. The amino acid adenylate component in each of these complexes is highly reactive but the continued binding of the enzyme prevents non-specific reactions. The same enzymes are involved in the following step also, namely,
2. The transfer of individual, activated amino acids to individual tRNA molecules. This step too is highly specific in that, while two or more varieties of tRNA may carry the same amino acid, a given type of tRNA molecule is confined to carrying a particular amino acid (see p. 162). Thus, the number of species of tRNA is not less than the number of amino acids involved in protein synthesis.
3. The individual attachment of 70S ribosomes[58] to mRNA. This involves the 30S sub-unit of the former and the 5′ terminal of the latter.
4. The alignment of AA-tRNA molecules along the length of mRNA templates.
5. The relative movement of the ribosomes along the mRNA molecules and their detachment on reaching the 3′ terminus. This movement involves ribosome/tRNA interaction. In fact, it appears that the accurate alignment and orientation of the charged tRNA on the

mRNA strand is itself sequential and depends on the ionic bonding of tRNA to high-affinity attachment sites on the 50S sub-unit of the ribosomes. The passage of the ribosome is accompanied by the progressive peptide linking of the amino acids of charged AA-tRNA molecules and the release of uncharged tRNA. The released tRNA can repeat its adaptor function and the exposed mRNA can accept further AA-tRNA molecules (see 7).

6. The progressive peeling-off of the polypeptide as polymerization proceeds. The growing polypeptide remains attached to the RNA complex at its growing end until its synthesis is complete. This end is a C-terminal so polypeptide synthesis must proceed from the N-terminal which, therefore, corresponds with the 5′ terminal of the mRNA.

7. The movement of the ribosome away from the 5′ end of mRNA allows the progressive attachment of other ribosomes so that a number of them may be simultaneously associated at different points along the length of the same messenger molecule (polyribosome complex).[72] Consequently, a number of incomplete polypeptides at different stages of sequential polymerization and, therefore, of different lengths may be attached to the same polyribosome complex at any one time (Fig. 7.6).

Much remains to be discovered about the detailed chemistry of the above sequences and though it may be inaccurate in its details, there is considerable experimental evidence in support of its main features (see also p. 164).

The question now is—How does the genetic material intercede in this synthetic sequence and does its epigenetic function end with the determination of primary protein structure?[9]

All three genera of non-genetic RNA, their species and varieties, are synthesized on DNA templates according to those same rules of complementary base pairing as apply, *mutatis mutandis*, during DNA replication. This process of DNA transcription proceeds in the 3′ → 5′ direction, i.e. RNA is synthesized from the 5′ to the 3′ terminal. Only one of the polynucleotide columns in a segment of duplex DNA is actually transcribed, the complementary chain being quiescent in this respect.[48]

Ribosomal and transfer RNA have extensive secondary and tertiary structures and their base sequences are not translated into AA sequences. Thus, the gene/polypeptide hypothesis discussed earlier is not relevant to those genes which code for these types of RNA.

The base sequence of mRNA, on the other hand, is translated and, subject to the qualifications given below, determines the primary structure of the resulting polypeptide. Translation involves the sequential

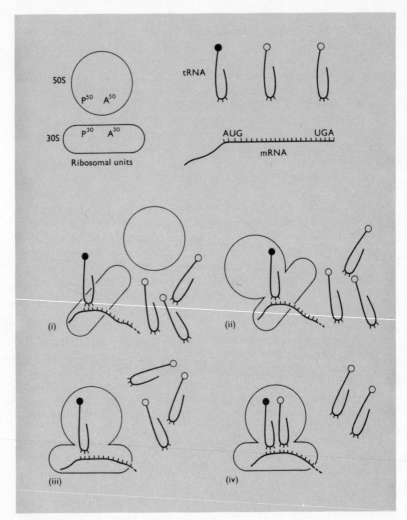

**Fig. 7.6** The principal components and possible steps in protein synthesis. (Top) Ribosomal sub-units with paired tRNA attachment sites (P and A). Charged tRNA with initiating (solid) and other (open) amino acids. In all probability there is no tRNA for stop signals. mRNA with initiating, interstitial and terminating codons. The initiating codon is probably preceded by a (purine rich?) region which is not translated. (i) Pairing between anticodon of N-formylmet-tRNA and initiating codon of mRNA, and binding to P-site of 30S sub-unit. (ii) Attachment of 50S sub-unit to initiating complex with the tRNA binding to its A-site. (iii) Clockwise displacement of 50S relative to 30S so that the initiating tRNA becomes associated with the P-sites of both 30S and 50S. (iv) Owing to (iii),

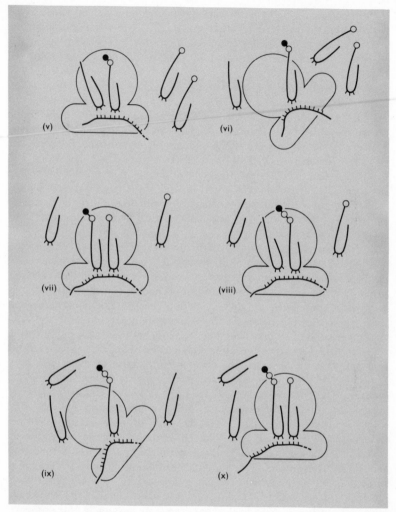

the second codon is aligned in relation to the A-sites of both 30S and 50S which, following codon–anticodon pairing, receive a second charged tRNA. (v) Transfer of initial AA to second AA by peptide bond formation. (vi) Release of discharged initiating tRNA. Anticlockwise rocking of ribosome so that the dipeptide tRNA becomes attached to the P-site of 30S, 50S A-bonding and codon–anticodon pairing is maintained. (vii) Clockwise rocking of ribosome so that the dipeptide tRNA becomes attached to P-site of 50S, 30S P-bonding and codon–anticodon pairing maintained. This moves the messenger along and sets the next codon in register. (viii–x) See iv–vi. The polypeptide is released on reaching a stop codon. 30S and 50S sub-units separate on leaving the template.

definition from the point of translation initiation of trinucleotide sequences (codons) on the mRNA. In other words, the three-letter codons of the genetic language are defined by a 'reading frame' which is set in register from the initiating point and not by punctuation marks between fixed codons.

Amino acids cannot align directly on mRNA, and tRNA plays an important adaptor role in this connection. Thus, the various species of AA-tRNA queue up along the mRNA template according to the codon sequence of the latter because mRNA/tRNA binding involves pairing between triplet codons of mRNA and complementary anticodon triplets of tRNA. From this it should be clear that the mRNA can be held to determine the final AA sequence only if there is a fixed relationship between the anticodon of the tRNA and the site recognized by the activating enzyme, because the recognition site determines the nature of the attached AA. Thus, the detailed structure of tRNA assumes considerable importance.

In fact, the complete primary structure of some of the tRNA varieties specific for alanine, serine and tyrosine has now been established. All tRNA is peculiar in containing a large number of unusual bases such as pseudouridine (U), dihydroxyuridine (DiHU), inosine (I), various methylated and even sulphur-containing bases. These are scattered throughout the molecule but they appear to be absent near the AA-attachment end. The unusual bases are probably important in regulating the secondary structure for while tRNA is essentially a single-stranded molecule of some 80 bases, the chain is folded back on itself (Fig. 7.7). However, the secondary structure is only partly helical and various tertiary structures are thus possible. Little is known about this level of organization but cation (especially $Mg^{++}$) concentration appears to be important in its stabilization. Thus, although tRNA is a polynucleotide, its structure and functions have much in common with those of enzymes.

Amino acids are accepted at the 3' terminal of tRNA and, following the usual convention of writing nucleotide sequences in the 5'-3' direction, the trinucleotide sequence adjacent to the AA terminal is CCA in all species of tRNA. Five serine-specific varieties of tRNA have been isolated from yeast and they have the same hexanucleotide sequence at the 3' terminal. But the valyl-oligonucleotides from yeast and rat RNA are different and the region recognized by the activating enzyme is not likely to be near the AA attachment terminal.

Further, although they are required to show corresponding specificities, the functions of codon recognition and enzyme recognition may be performed by different or overlapping parts of the tRNA molecule. Certainly, the two functions can be affected to different extents by particular treatments and mutation may affect them independently.

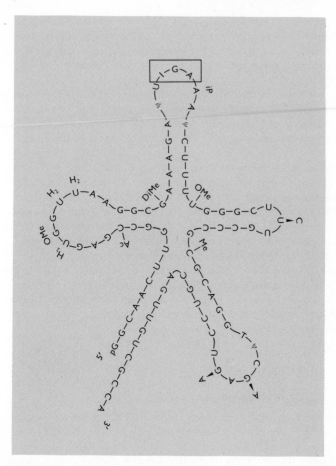

**Fig. 7.7** The base sequence and possible secondary structure of serine tRNA. Two varieties of this species of tRNA were found to differ at three base locations, indicated by the solid triangles. (From Zachau, H. G. *et al.* (1966), *Cold Spring Harb. Symp. quant. Biol.*, **31**, 417–24.)

It is also possible that the activating enzyme does not recognize a particular site but rather a particular conformation created by the co-operation of several regions and compatible with a restricted range of primary structures. Different varieties of the same species of tRNA may be codon-specific in some cases or at least show preference for some codon synonyms. But the known varieties of tRNA which are specific for a given AA are not always as numerous as the codons assigned to the

AA concerned. Therefore, a given tRNA molecule may be able to recognize more than one synonym (see p. 168). Conversely, in cases such as the gly-tRNA of yeast, there appear to be more varieties than are minimally required and a given codon may be recognized by more than one tRNA variety.

The situation in relation to the activating enzymes may be similar to some extent. For example, it is likely that a single enzyme charges the 4 or 5 varieties of leucine-tRNA which are known in *E. coli*. But the analogous enzyme from yeast can charge only two of the *E. coli* tRNA varieties.

Studies on unfractionated tRNA from *E. coli*, yeast and rat suggest that a GTψCG sequence may be common to all tRNA molecules and that it could be concerned with the AA-non-specific binding between tRNA and the 50S ribosomal unit. The 3′ terminal CCA sequence is important in this connection also for while both charged and uncharged tRNA can bind to ribosomes, molecules lacking the CCA sequence cannot.

Scientific progress is rarely linear and the studies which led to the scheme outlined above form an interlocking, mutually supporting series which cannot be discussed in detail. However, the principal experiments underlying the general scheme will be given some attention before the details of the genetic code and the more debatable aspects of information transfer are discussed.

## THE GENETIC CODE[75]

### Single-strand transcription

Mutations in structural genes which lead to amino acid substitution affect the primary structure of only one polypeptide chain. This suggests that only one mRNA molecule is produced per DNA duplex.[48] This could mean that the production of a messenger is the joint responsibility of the two complementary DNA chains. But annealing experiments have shown that the two-strand/one-messenger relationship depends on the fact that only one of the two DNA chains is transcribed *in vivo*.

Thus, when the two columns of duplex DNA are separated by heating (see p. 23) and subsequent cooling is allowed to occur in the presence of single-stranded mRNA transcribed by homologous DNA *in vivo*, duplex molecules of two kinds are formed, namely, renatured DNA and DNA/RNA hybrids. In the case of SP8 and the double-stranded replicating form of ϕX174 it has been shown that, over a given region, only one of the two complementary columns of the DNA duplex participates in the formation of DNA/RNA hybrid molecules. This demonstration is simpler in SP8 because its denatured DNA consists of a heavy strand rich in pyrimidines and a light strand rich in the complementary purines. The strand difference is large enough to allow their centrifugational and chromatographic separation and only the pyrimidine-rich segments hybridize with homologous RNA synthesized in an infected host.

In the smaller viruses the transcribed strand may be longitudinally continuous throughout the genome but this is not a necessary state. For example, we have seen that, owing to the opposite polarity of the two chains in duplex DNA, inversion involves strand exchange (see p. 21). But inversion of the $lac^+$ region in *E. coli* does not interfere with its function. Of course, the transcription of opposite strands over different regions means that they are read in opposite physical directions because in both cases the DNA is transcribed $3' \rightarrow 5'$. Further, certain units of transcription (see p. 188) of *E. coli* are inverted relative to their counterparts in *Salmonella*, which otherwise has a very similar linkage map.[30]

Opposite orientation is evident also within the genome of $T_4$ phage. For example, the lysozyme and head protein genes of this virus are translated in opposite directions relative to their orientation in the genetic map.

The basis of *in vivo* strand selection is not clear because in many *in vitro* systems both strands of duplex DNA are transcribed. It has been suggested that pyrimidine-rich clusters, perhaps by modifying the secondary structure of DNA, may serve as recognition sites for the initial attachment of RNA polymerase. They could thus determine both the choice of strand and, thereby, the physical direction of transcription.

In fact, pyrimidine-rich clusters (especially C sequences) have been found in DNA from various organisms and they are often asymmetrically distributed between the two strands. Indeed, in the small coliphages, $T_3$ and $T_7$, they appear to be confined to one strand (compare SP8).

## Non-overlapping code

A genetic language in which one nucleic acid base specified one amino acid would allow the coding of only four kinds of AA. A coding ratio of two, on the other hand, would permit the specification of sixteen AA because this is the number $(4^2)$ of possible pairs into which four bases can be arranged when order is taken into account. But since 20 kinds of AA participate in protein production, a coding ratio of at least three has long been recognized as the minimum.

A coding ratio greater than one introduces the possibility of an overlapping code in which a given base location can form part of more than one code word. In this event, single base substitutions could lead to changes in 2 or more neighbouring AA according to the word size (i.e. the coding ratio). Further, constraints would be imposed which would prohibit certain AA sequences. However, as early as 1956 it was clear that the AA sequence of insulin was incompatible with such constraints and the same is true for subsequently-determined AA sequences. Also, the numerous examples of single AA replacements, now known in mutant proteins of TMV, tryptophan synthetase, alkaline phosphatase and haemoglobin, argue against an overlapping code.[13] So too does the observation that polypeptides containing only one type of AA (homopolypeptides) can be synthesized *in vitro* under the direction of synthetic ribopolynucleotides consisting of repeating triplet sequences (e.g. AAGAAG—). This observation also indicates a coding ratio of three, but the earliest evidence of codon size came from a study of mutations induced with acridine.[4]

## The triplet code and the reading frame

There are reasons for believing that the mutagenic effects of acridine and related compounds depend on the addition (+) or deletion (−) of single bases during DNA replication. When two, independently-derived acridine mutants are crossed, some double mutants arise as a result of crossing-over between the two mutant sites. These are expected to give mutant phenotypes and, when the original mutations involve different genes, this expectation is invariably fulfilled.

However, when two mutants which are defective in the same polypeptide are involved, certain double acridine mutants resemble wild-type much more closely than either of the single mutants. The members of such mutually-suppressing pairs are arbitrarily designated plus and minus (Fig. 7.8). By reference to an arbitrarily chosen standard pair a

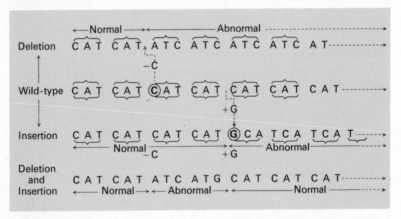

**Fig. 7.8** The mutually-suppressing effects of single base deletion (−) and insertion (+) mutants. The reading frames are defined from the left and the wild-type frame spells CAT. Deleted and inserted bases are shown in white circles.

consistent designation of other mutations of the same gene proves to be possible.[22] Thus, neither double plus nor double minus mutations approach wild-type. But plus-minus double mutants tend to show a degree of compensation which is roughly proportional to the distance between the mutant sites. Triple minus and triple plus mutants approach wild-type also, while other triple combinations are as defective as the single mutants irrespective of the mutant sequence (e.g. + + −, + − +).

In the case of phage lysozyme it has been shown that the protein produced by various mutually-suppressing combinations (+ −, − − − and + + +) differs from wild-type only with regard to short amino acid sequences, the precise length and location of which depend on the sites of mutation. More particularly, the abnormal segment corresponds with the inclusive intercept between the codons in which the two, plus and minus, mutations occurred, or, where three mutations of the same sign are concerned, with the region between the outermost of the codons involved (Fig. 7.9).

These results support the suggestion originally made by Crick and his co-workers, on the basis of studies on the r II region of T₄, that the genetic message is read as a series of 3-letter words which are defined sequentially from a fixed point.[77] On this view, the deletion or addition

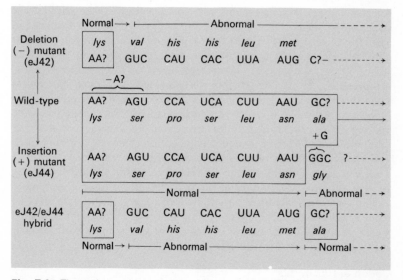

**Fig. 7.9** The codon and amino acid changes determined by frame-shift mutations in lysozyme of T₄ phage. (Data of Terzaghi, E. *et al.* (1966), *Proc. natn. Acad. Sci. U.S.A.*, **56**, 500–7.) Wild-type codons and amino acids are shown in white boxes.

of a base would change the spelling of *all* the words distal to the site of mutation. Consequently, the AA composition of the protein produced would resemble wild-type only over the region coded by the intercept between the sites of initiation and mutation. However, a plus mutation would correct the frame-shift resulting from a proximal minus mutation (and vice versa) and only the AA coded by the intercept between them would be affected. If this sequence is short or comparatively unimportant, the resulting protein is expected to be essentially wild-type in function.

Frame-shift mutations of this type are thus among the exceptions referred to earlier (p. 158) for they have a unidirectional spreading effect on the framing of codons and so they affect AA other than those specified by the codons in which they occur (see also p. 176).

A total of 64 ($=4^3$) three-letter codons can be formed on the basis of a four-letter alphabet—many more than the minimum number required. This suggests two possibilities.

First, the code is highly redundant. In other words, many of the possible triplets are 'nonsense' and do not code for any AA. In fact, from the 64 possible words, groups of twenty, one for each AA, can be selected so that, irrespective of the order in which the various codons of a group are placed, none of the overlapping triplets coincides with any of the selected codons. Various groups with this property can be selected but they cannot contain more than 20 words. Actually a comma-free code of this type was proposed to meet the problem of internal punctuation. But although such a code is aesthetically pleasing, experiment shows that nature has been less elegant if more cautious (see p. 171). For example, if only 20 or so of the 64 possible words made sense, frame-shift mutations should generate many nonsense triplets. But studies on acridine-induced mutants showed that very little nonsense was created when the reading frame was changed. Further, ribopolynucleotides containing randomly linked bases can direct polypeptide synthesis *in vitro* which again indicates that the code is not over-burdened with nonsense.

Of course, the demonstration that the reading frame is set in register from an initiating point obviates the requirement for internal commas.

The second possibility is that the code is highly degenerate, that is, many of the code words are synonymous and code for the same AA. This is now known to be the case as shown by the biochemical studies which led to the cracking of the genetic code.

## Codon assignments

Biochemical approaches of two main kinds have been used in attempts to assign particular code words to the twenty amino acids which participate in protein synthesis.

First, *in vitro* studies have been made on the incorporation of amino acids in cell-free systems containing synthetic ribopolynucleotides.[68] In the early experiments of this type, codon assignments were made on the assumption that the bases in the artificial messengers were arranged at random and in proportion to their relative concentration in the reaction mixtures used in their synthesis. Given these assumptions, the relative frequencies of the theoretically-possible three-letter codons could be calculated and compared in kind and quantity with the AA whose incorporation was stimulated by the ribopolynucleotides concerned. The order of the letters in the code words could not be determined on this basis, but subsequently polymers of repeating di-, tri-, and tetra-nucleotides of known sequence were used. In the production of these *in vitro* systems the protein-synthesizing components of broken cells (usually *E. coli*) are separated from the remainder by gentle centrifugation. During these manipulations natural mRNA is subject to mechanical and enzymic break-down. Artificial messengers can then be introduced. Many of these have been prepared with the aid of a polynucleotide phosphorylase obtained from *Azotobacter vinelandii* or *Micrococcus lysodeikticus*. This enzyme differs from RNA polymerase in not requiring the presence of DNA as a primer or template for the synthesis of RNA.

The second approach involved a study of the effect of various ribotrinucleotides of known sequence on the stimulation of binding between specific AA-tRNA molecules and ribosomes.

The results obtained by these two methods are in essential agreement and are summarized in Table 7.6. The degeneracy of the code is obvious but it is not without restriction and pattern. Thus, nine of the AA are each represented by two code words. In all of these the degeneracy depends on variation in only the third letter and even this is restricted to a choice between the two pyrimidines (U or C) or the two purines (A or G). One AA has 3 and five have 4 synonyms and here too only the third letter is variable. *Leu* and *arg* each have six synonyms in which the second letter is constant and only in the six codons for serine is variation found in respect of all three letters. Three of the codons, amber, ochre and UGA, are nonsense (see p. 173).

This mRNA/AA dictionary was compiled on the basis of *in vitro* studies using cell-free fractions of bacteria. To what extent, therefore, is it applicable to *in vivo* synthesis and does it hold for all organisms?

## Amino acid replacements

The correlation between the accepted codon assignments and single AA substitutions is well illustrated by the A protein of tryptophan synthetase (Fig. 7.10). Thus a series of progressive and reverse mutations affecting a glycine site in peptide TP3(I) of this molecule can be

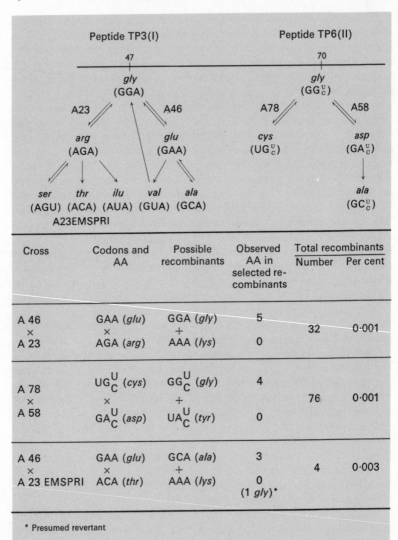

**Fig. 7.10.** Correlation between accepted code words and serial AA replacements following mutation and reversion at *gly* sites, 47 and 70 (see Fig. 7.5), in adjacent tryptic polypeptides of tryptophan synthetase in *E. coli*. (Data of Guest, J. R. and Yanofsky, C. (1966), *Nature, Lond.*, **210**, 799–802.) The two mutations A 23 and A 46 cause the *gly* of TP3(1) to be replaced by *arg* and *glu*, respectively, and they evidently affect adjacent bases in the same codon. Even so they can be recombined at the rate of 0·002 % to restore the wild-type sequence.

explained on the basis of single base substitutions and the codon assignments given in Table 7.6. Five other mutations involving a glycine site of peptide TP6(II) of the same molecule can be similarly explained but only if the original *gly* codon is taken as GG$_C^U$ for this site as opposed to GGA for the previous one. Apart from supporting the codon assignments, these results indicate the occurrence of degeneracy within a single gene.

**Table 7.6** The genetic dictionary. The trinucleotide codons are written in the 5′–3′ direction. Hence the first base represents the 5′ terminal and the third base the 3′ terminal. Substitutions of amino acids occupying the same column, row or block can be determined by single base changes in the first, second or third letters of the codon, respectively

| First base | Second base | | | | Third base |
|---|---|---|---|---|---|
| | U | C | A | G | |
| U | UUU UUC *Phe* <br> UUA UUG *Leu* | UCU UCC UCA UCG *Ser* | UAU UAC *Tyr* <br> UAA UAG *Stop* † | UGU UGC *Cys* <br> UGA *Stop* <br> UGG *Try* | U C A G |
| C | CUU CUC CUA CUG *Leu* | CCU CCC CCA CCG *Pro* | CAU CAC *His* <br> CAA CAG *Gln* | CGU CGC CGA CGG *Arg* | U C A G |
| A | AUU AUC *Ile* <br> AUA <br> AUG *Met* * | ACU ACC ACA ACG *Thr* | AAU AAC *Asn* <br> AAA AAG *Lys* | AGU AGC *Ser* <br> AGA AGG *Arg* | U C A G |
| G | GUU GUC GUA GUG *Val* | GCU GCC GCA GCG *Ala* | GAU GAC *Asp* <br> GAA GAG *Glu* | GGU GGC GGA GGG *Gly* | U C A G |

* AUG = Met or Chain Initiation.
† UAA (Ochre) and UAG (Amber) are suppressible nonsense codons.

It is of particular interest that the spontaneous AA replacements known in human haemoglobin can be correlated with single base substitutions in codons which were assigned on the basis of *in vitro* studies with bacterial systems. This is true also for nitrous oxide induced mutations affecting the coat protein of TMV. In this case, over 85% of known mutations can be explained on the predicted basis of nitrous oxide mutagenesis, namely, A to G and C to U transitions. The remainder involve alternative transitions or transversion (see p. 46).

In fact, only one of the 70 or so single amino acid substitutions which are known in various proteins cannot be explained on the basis of single base substitution in assigned codons. The exceptional case is an *asn* → *arg* substitution in a spontaneous mutant of TMV.

## Chain initiation and termination

Homotriplet sequences of three kinds can be defined in a ribopoly-nucleotide of the poly AAG type (viz. AAG-AAG etc., AGA-AGA etc. and GAA-GAA etc.) according to the point from which codons are defined. *In vitro*, this polymer can direct the synthesis of homopoly-peptides of three kinds, namely, polylys, polyarg and polyglu. This result supports the triplet nature of the code and provides information on codon assignments. But it also indicates that translation can be initiated at different points along artificial messengers.

Further, the AAA····AAC polynucleotide directs the synthesis of *lys* polypeptides with *asN* at the —COOH terminal as well as homolys poly-peptides of various lengths. Results of this type too are important in relation to codon assignment and they also indicate the direction of reading. But, in addition, they show that translation can be terminated as well as initiated at various points.

Clearly, neither of these situations can be tolerated *in vivo* and they would be avoided if a specific initiating site occurred at the 5' end and a terminating signal of the 3' end of the translated portion of natural messengers.

### Initiation

In a cell-free system from *E. coli*, natural RNA from $f_2$ or R17 phage directs the synthesis of coat protein which has *N-formylmet* at the amino terminal followed by -*ala-ser-asN-phe*. *In vivo*, however, this protein has *ala* at the amino terminal followed by -*ser-asN-phe*. It has been suggested, therefore, that the coat protein is actually synthesized in the same way *in vivo* and *in vitro* but that the amino terminal *N-formylmet* is enzymic-ally split off in the former. Other proteins may be initiated in the same way but the extent of the splitting off may vary. It is significant in this connection that the amino terminal is occupied by *met* or *ala* in 85% of the soluble proteins of *E. coli* while *ser* or *thre* occur at this site in 13% of them.

The importance of *N-formylmet* is further indicated by the fact that *in vitro* protein synthesis directed by artificial or certain natural mes-sengers is dependent on it when the $Mg^{++}$ concentration is low.

Now, *met* is one of the two AA for which there is no codon degeneracy and, clearly, it would be intolerable if the unique *met* codon, AUG, served as an initiator wherever it occurred. In fact, this difficulty is intensified by the finding that, in binding assays, N-formylmet-tRNA, unlike *met*-tRNA, will respond not only to AUG but to GUG, UUG and even CUG to some extent. Studies on AA incorporation support the view that both AUG and GUG and even GUA can serve as initiators. For instance, at low $Mg^{++}$ concentration and in the presence of N-formylmet-tRNA, polyGU directs the synthesis of a *cys-val* co-polypeptide with *N-formylmet-cys-val* at the amino terminal. This suggests the following framing and coding:

$$\underbrace{G\ U\ G}\ \underbrace{U\ G\ U}\ \underbrace{G\ U\ G}$$
$$N\text{-}formylmet\ \ -\ \ cys\ \ -\ \ val$$

Thus, there appears to be an element of ambiguity in that GUG, in this instance, can code both for *val* and *N-formylmet* depending on its position in the message. It has been suggested, therefore, that certain triplets which normally code for other AA actually specify *N-formylmet* when they occur at the $5'$ end of a translated section and thus serve as initiators and setters of the reading frame (see also p. 164).

In fact, there is evidence that the tRNA which carries *N-formylmet* ($tRNA_F$) is structurally different from that which transfers *met* ($tRNA_M$). A basis for discrimination thus exists which may be related to binding with the ribosome. It must be stressed that while the role of *N-formylmet* has been intensively studied in *E. coli* it may not be unique in respect of the chain initiating function.[1]

*Termination*

Certain (amber) mutants of $T_4$ produce only amino terminal fragments of the head protein when they are grown on so-called non-permissive or restrictive hosts ($su^-$). The fragments vary in length with the position of the mutation concerned (Table 7.7). Thus, if three mutations of this type map in the sequence ABC, the B-fragment is intermediate in length. This suggests that mutations of sense–nonsense type are involved so that the sequential translation of the mRNA is prematurely terminated at a site corresponding to that of the mutant codon (but see p. 176). Comparable mutations are known in other genes, other organisms, and for one such mutant in $f_2$ phage it has been shown that *in vitro* translation is similarly interrupted.

Mutants of this kind are known to revert but exact back-mutation is not always involved. For instance, a chain terminating mutation affecting the alkaline phosphatase of *E. coli* can be referred to an interstitial codon

**Table 7.7** Amino fragments and chain-terminating amber mutants. Certain peptides which are characteristic of the head protein of wild-type T4D phage are absent from amber mutants. The correlation between the peptide deficiencies and the sites of the various amber mutations are consistent with the view that polypeptide synthesis is terminated at different points according to the map location of the amber mutant. (Data of Sarabhai, A. S. *et al.* (1964) *Nature, Lond.*, **201**, 13–17.)

| Linkage map | B17 | | B272 | | H32 | | B278 | | C137 | | H36 | | A489 | | C208 |
|---|---|---|---|---|---|---|---|---|---|---|---|---|---|---|---|
| | | 1·15 | | 0·5 | | 1·82 | | 0·33 | | 0·22 | | 1·04 | | 0·34 | |
| Peptides | Cys | | HisT7c | | TyrC12b | | TryT6 | | ProT2a | | TryT2 | | TryC2 | | HisC6 |
| Wild-type | + | | + | | + | | + | | + | | + | | + | | + |
| C208 | + | | + | | + | | + | | + | | + | | + | | − |
| A489 | + | | + | | + | | + | | + | | + | | − | | − |
| H36 | + | | + | | + | | + | | + | | − | | − | | − |
| C137 | + | | + | | + | | + | | − | | − | | − | | − |
| B278 | + | | + | | + | | + | | − | | − | | − | | − |
| H32 | + | | + | | + | | − | | − | | − | | − | | − |
| B272 | + | | − | | − | | − | | − | | − | | − | | − |
| B17 | − | | − | | − | | − | | − | | − | | − | | − |

Direction of translation  NH₂ ⟶ COOH

which specifies *try* in the wild-type enzyme. But in various revertants which produce complete chains, this AA site can be occupied by *glN*, *glu*, *tyr*, *ser*, *leu* or *lys* as well as the original *try*.

The original nonsense mutation and all these reversions can be explained by single base substitutions if nonsense is assigned to the UAG codon (Table 7.8).

**Table 7.8** A chain terminating (amber) mutation affecting alkaline phosphatase in *E. coli* and the nature of the secondary, complete chain, mutants derived from it. Note that the observed substitutions occur in the same column, row or block of the dictionary (Table 7.6) and can therefore be ascribed to alterations in the first, second or third letter of the codon, respectively. The altered base is shown in heavy type, transitions in roman and transversions in italic. (Data of Weigert, M. G. and Garen, A. (1965) *Nature, Lond.*, **206**, 992–4.)

| Strains | Amino acid substitutions | Probable, single-letter codon changes |
|---|---|---|
| Wild-type<br>Interstitial AA | Try | UGG |
| Primary amber mutant<br>Interstitial chain termination | COOH terminal | UAG |
| Back mutation of amber<br>Interstitial AA | Try | U**G**G |
| Reversions of amber<br>(i)<br>(ii)<br>(iii)<br>(iv)<br>(v)<br>(vi) | Glu<br>GIN<br>Lys ?<br>Ser<br>Leu<br>Tyr | *G*AG<br>**C**AG<br>*A*AG<br>U**C**G<br>UU*U*G<br>UA*U*/*C* |

Comparable studies on another class of mutants (Ochre) which similarly produce incomplete chains show that nonsense can be ascribed to the UAA codon also. Further, binding assays, AA incorporation studies and genetic considerations indicate UGA as a third word which does not code for any AA. But the extent to which these three nonsense codons function as natural terminators is not clear.

It will be appreciated that base substitution is not the only way in which a sense–nonsense mutation can arise. Thus, a frame-shift mutation, by altering the content of code words, can generate nonsense codons distally as follows:

$$\text{Lys - Lys - AsN - Arg - Lys} \longrightarrow$$
$$\overbrace{A\ A}\ \overbrace{A\ A\ A}\ \overbrace{A\ A\ U}\ \overbrace{A\ G\ A}\ \overbrace{A\ A\ A}$$

Insert | A

$$A\ A\ A\ A\ A\ A\ A\ A\ U\ A\ G\ A\ A\ A$$
$$\underbrace{\text{Lys}}\ \text{-}\ \underbrace{\text{Lys}}\ \text{-}\ \underbrace{\text{Lys}}\ \underbrace{\text{Nonsense}}$$

In this event, the sites of mutation and chain termination will not coincide but the interval between them will be represented by a modified AA sequence. It will be appreciated also that nonsense codons, by causing the premature termination of translation, will necessarily affect amino acid sites distal to them. Thus, unlike missense mutations, both frame-shift and nonsense mutations have a unidirectional effect on AA sites other than those specified by the codons in which they occur.

## Ambiguity

The first step in deciphering the genetic code was the demonstration that polyU directed the incorporation of *phe*. But, *in vitro*, it also codes for *leu* in small, but significant, amounts. Further, each of a number of sites in the α-chain of haemoglobin from a single individual rabbit can be occupied by different AA. For example, position 48 may be equally occupied by *leu* or *phe*.

Results such as these suggest that a given codon may be translated in more than one way, according to a variety of misrecognitions. Thus, changes in the primary, secondary or tertiary structure of either tRNA or rRNA could lead to a distortion of codon-anticodon interaction. Alternatively, an erratic RNA polymerase may mistranscribe the DNA, or an aberrant aminoacyl tRNA synthetase, by misrecognizing either the AA or the tRNA may combine the wrong amino acid with the right adaptor. In these events the codon–anticodon relationship would remain unaffected. All these changes could be determined by mutation or else they could be mere modifications caused by the local environment (see also p. 172).

Different varieties of tRNA specific for the same AA may recognize different codon synonyms preferentially if not specifically. But a given variety of tRNA can recognize more than one codon synonym.[21] For example, a highly purified species of yeast ala-tRNA can recognize three *ala* codons, GCU, GCC and GCA. This capacity for multiple codon recognition means that the number of tRNA varieties for a given AA need not be as large as the number of corresponding codon synonyms.

However, in some cases (e.g. yeast gly-tRNA) it is probable that there are more tRNA varieties than are minimally required.

It was pointed out earlier that most of the degeneracy in the genetic code involves the third letter of the codon and this applies also to multiple codon recognition. Thus, in the above example in α haemoglobin, the codons for *leu* and *phe* can differ by only one letter and this letter can be the third (see also p. 169).

From the above discussions it is clear that even the production of an immediate gene product is a venture in which the products of many other genes co-operate. There are, therefore, many areas in the epigenetic landscape where genes can influence each other's immediate and ultimate activities. And it is to a consideration of these interactions that we now turn.

# 8

# *Epigenetic Interactions*

Structural genes specify mRNA and, hence, they can be held to determine the primary structure of polypeptides. But, clearly, polypeptide production itself depends on the co-operative interaction of genes of different kinds.[90] Indeed, the role of structural genes in determining primary protein structure can be maintained only if there is a consistent relationship between the anticodons and the amino acid recognition sites of tRNA (see p. 162).

Less immediately, the enzymic products of different structural genes themselves co-operate in creating variously interconnected metabolic pathways and further interactions occur between the immediate and end products of these pathways (Table 8.1).

It is clear, therefore, that while character differences, even those in the

**Table 8.1** Gene interaction in *Pisum sativum*. The difference between tall and short lines in Mendel's classical experiments was determined by a single gene difference. This does not mean the height in peas is determined by a single gene. In fact, seven loci with major effect are known. (Data of Lamprecht, H. (1962) *Agri Hort. Genet.*, **20**, 23–62.)

| Chromosome No. | I | II | V | VI | VII |
|---|---|---|---|---|---|
| 1. Genes influencing length of internodes | Cot | Coe | Coh | Cona | — |
| 2. Genes influencing number of internodes | Mie | — | Miu | — | Min |

peripheral phenotype, may be attributable to single and simple genetic differences, no aspect of the phenotype itself can be attributed to any one genetic element. We do not intend to discuss peripheral pleiotropic effects and interactions in any detail but some attention will be paid to their underlying molecular basis.

## INTERACTIONS BETWEEN STRUCTURAL GENE PRODUCTS

Auxotrophic mutants are detected and isolated owing to their inability to grow on minimal medium. However, when some of them are plated together, growth is observed even on unsupplemented medium. This is true, for example, of arginine- and leucine-requiring mutants in *E. coli*. This mutual supplementation or syntrophism depends on the ability of each strain to synthesize and excrete into the medium the amino acid required by the other. It does not, therefore, depend on cell fusion much less on recombination between the parental genomes.

Thus, the prospects for cross-feeding between sympatric auxotrophs are limited by the functional and, hence, mutational relationship between them and also by the nature of the required substances, their lability, diffusibility, etc. For example, L-histidinol is a sufficient supplement for certain histidineless mutants of bacteria but it cannot normally serve this function in *Neurospora* for reasons of permeability. But given that the required metabolites are released and can be taken up, the pattern of cross-feeding can indicate the functional relationship between the affected genes.

For example, when, as in the above example, two mutants are defective in different unrelated pathways, different enzyme functions being affected, they are expected to be mutually syntrophic. On the other hand, when two mutants are blocked at different stages of the same pathway, neither can produce the end product but cross-feeding remains a possibility though it cannot be immediately mutual. For example, if two mutants, X and Y, are blocked in the A-E pathway as follows:

| Mutants | Pathway |
|---------|---------|
| X       | A $\longrightarrow$ B $\longrightarrow$ C—$\mid\mid$-→ D -----→ E |
| Y       | A $\longrightarrow$ B—$\mid\mid$-→ C-----→ D -----→ E |

Y, which is blocked at an earlier stage, has nothing immediately to contribute to X which X cannot make for itself. But X can supply Y with C which it can then convert to end product E. However, if as a result of this subsidy, Y produces E (or D) in excess, it may leak into the medium and satisfy X. Therefore, the pattern of cross-feeding can provide some information on the sequence of the reactions in which the genes are indirectly involved.

Clearly, the interactions which enable two blacks to make a white across the cell membranes which separate them, should be facilitated and even extended by the breakdown of those barriers. This is achieved in various degrees by the synthesis of heterokaryons, heterogenotes, heterozygotes or mixedly infected bacteria.

As expected, recessive mutants which are defective in different pathways or different stages of the same pathway generally yield hybrids which closely approach the wild phenotype though the dominance of the corresponding wild-type genes may not be as complete as peripheral observation suggests.

At first sight, complementary interactions of this kind would seem to be prohibited when the two mutants are blocked at one and the same step and defective for one and the same enzyme. But this is not so. In fact, such complementation may be the rule or the exception depending on the nature of the enzyme and the molecular basis of mutual dependence.

## Intercistronic compensation

Certain proteins, including some with enzymic functions, have a heteromultimeric quaternary structure (p. 148). Thus, the tryptophan synthetase of *E. coli* contains A and B units, the primary structures of which are determined by distinct but adjacent segments of the genome.

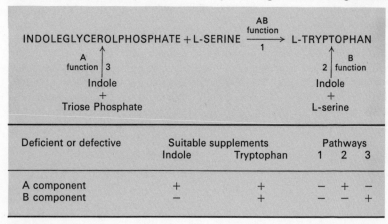

| Deficient or defective | Suitable supplements | | Pathways | | |
| | Indole | Tryptophan | 1 | 2 | 3 |
| --- | --- | --- | --- | --- | --- |
| A component | + | + | − | + | − |
| B component | − | + | − | − | + |

**Fig. 8.1** The reactions catalysed by tryptophan synthetase and the limited capacities of its A and B components in *E. coli*. Note that in its catalytic capacities the AB heteromultimer is greater than the sum of its parts.

This enzyme can catalyse three steps in the terminal stages of tryptophan synthesis, but separate and purified A and B components, like B-defective and A-defective mutants, have only limited capacities and even these are enhanced by combination (Fig. 8.1). What is more, the wild phenotype is restored on crossing A and B mutants.

In the situations considered so far the recessive mutations concerned were in different structural genes in that they affected the primary structure of different polypeptides. Consequently, the hybrids produced by mutant$_1$ × mutant$_2$ crosses could reasonably be expected to contain a full complement of normal proteins (as well as defective ones) and, hence, to approach the wild-type. But this expectation does not apply when mutations of the same structural gene are involved. In fact, in the case of tryptophan synthetase, pairs of A mutants on the one hand, and pairs of B mutants, on the other, never yield wild-type hybrids. But these hybrids can give wild-type recombinants showing that strictly (structurally) allelic mutations need not be involved. In fact, the presence versus absence of complementation can be used as a test of functional allelism, closely linked mutant sites which fail to complement being provisionally

| Dihybrid genotypes | Possible phenotypes | |
| --- | --- | --- |
| | One mutation in each of two non-homologous cistrons | Two mutations in the same or homologous cistrons |
| Trans $\dfrac{m_1 \quad +}{+ \quad m_2}$ | Wild-type | Mutant |
| Cis $\dfrac{m_1 \quad m_2}{+ \quad +}$ | Wild-type | Wild-type |
| Conclusions | $\underbrace{m_1}\;|\;\underbrace{m_2}$ two cistrons | $\underbrace{m_1 \quad m_2}$ one cistron |

Fig. 8.2  The cis-trans test for functional allelism. Cis-double heterozygotes are expected to be wild-type whatever the locations of the mutations provided they are recessive. When pairs of mutations occupy different cistrons, the trans dihybrid also contains a full complement of wild-type cistrons. But trans dihybrids for mutations in the same cistron do not have that cistron in the wild-type state. Consequently, they are expected to be mutant.

allocated to the same functional unit or cistron (Fig. 8.2). However, the cis-trans test sometimes yields superficially ambiguous results.

## Intracistronic complementation

If both $m_1$ and $m_2$ fail to complement $m_3$ and gross aberrations can be ruled out, all three mutations can be allocated to the same cistron. On this basis, $m_1$ and $m_2$ are not expected to complement each other. But they sometimes do.

Under these circumstances, however, complementing mutants can still be classified together and regarded as intracistronic because they are connected indirectly by a continuous series of non-complementing mutations which are in the majority.

The complementation relationships of intracistronic mutations can be summarized diagrammatically in the form of a map in which:

1. Mutually complementing mutations are represented by non-overlapping lines.

2. Non-complementary mutations are represented by overlapping lines.

In most cases, the consistent application of these rules is compatible with the construction of a simple, two-dimensional map in which each mutant is represented by a single continuous line (Fig. 8.3). But complementation maps tend to become complicated (e.g. connected circles) when very large numbers of mutations are considered.

Intracistronic complementation has been shown to have the following general properties:

1. Although the peripheral phenotype of the hybrid may be wild-type, the complementation is never complete. Thus, there is always a quantita-

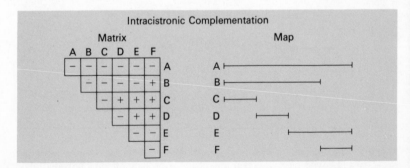

**Fig. 8.3** The phenotypes of the hybrids produced on crossing mutants (A-F) with recessive mutations in homologous cistrons and the map summarizing their complementation relationship (hypothetical). Plus = (Pseudo) wild-type, Minus = Mutant.

tive difference between the cis and trans dihybrids (cf. intercistronic compensation). The contradiction between the peripheral and underlying phenotype arises because low levels of enzyme activity are often sufficient to support the wild phenotype. For example, less than 5% of wild-type argininosuccinase activity is sufficient to confer arginine independence in *Neurospora*.

2. Different levels of complementation are shown by different mutants within the series. With rare exceptions, complementing trans hybrids usually give less, generally much less, than 25% of wild-type activity.

3. The enzyme produced by the trans hybrid is qualitatively different from wild-type, e.g. in its thermostability.

4. Intracistronic complementation is shown by only a minority of mutants, e.g. 12% in the case of the ad-3B mutants of *Neurospora*.

5. It is not shown by frame-shift or nonsense mutants but only by those which produce a protein with sufficient serological similarity to wild-type to react with antibodies prepared against the latter (cross reacting material —CRM).

The precise molecular basis of intracistronic complementation has not been elucidated and various possibilities can be entertained.[23]

For example, the tryptophan synthetase of *Neurospora*, unlike that of enteric bacteria, is not heteromultimeric but a single molecule which nevertheless shows the multifunctional properties of its bacterial counterpart. What is more, A-type and B-type functions (see p. 180) can be independently affected by mutation. Consequently, hybrids between mutants lacking different functions might be expected to show a simple additive effect comparable with that in intercistronic interaction, and thus to approach wild-type more closely than either mutant parent. However, there is no evidence for such an effect in any case of intracistronic complementation and it can be clearly ruled out in others, including the tryptophan synthetase of *Neurospora*.

Interallelic interaction can occur even when the interacting alleles are confined to separate nuclei in a heterokaryon. Direct interaction between genes is thus excluded. It has been suggested, therefore, that the restoration of normal enzyme activity depends on recombination between the mRNA molecules of the interacting alleles or even between their polypeptide products. But the evidence is against these views.[38]

The current hypothesis proposes that a homomultimeric quaternary organization provides the basis for intracistronic complementation. In these terms, the functional wild-type enzyme is regarded as a homomultimer consisting of two or more identical ('homologous') polypeptides. Mutations, by altering the primary structure of the polypeptide, may affect the secondary and tertiary configurations so as to prohibit the formation of stable multimers with the surface properties of the wild-type. However, the polypeptides produced by certain pairs of mutants may be modified in complementary ways so that hybrids between them produce quaternary structures which resemble the wild-type homo-multimer (Fig. 8.4). Of course, the hybrids are expected to produce the

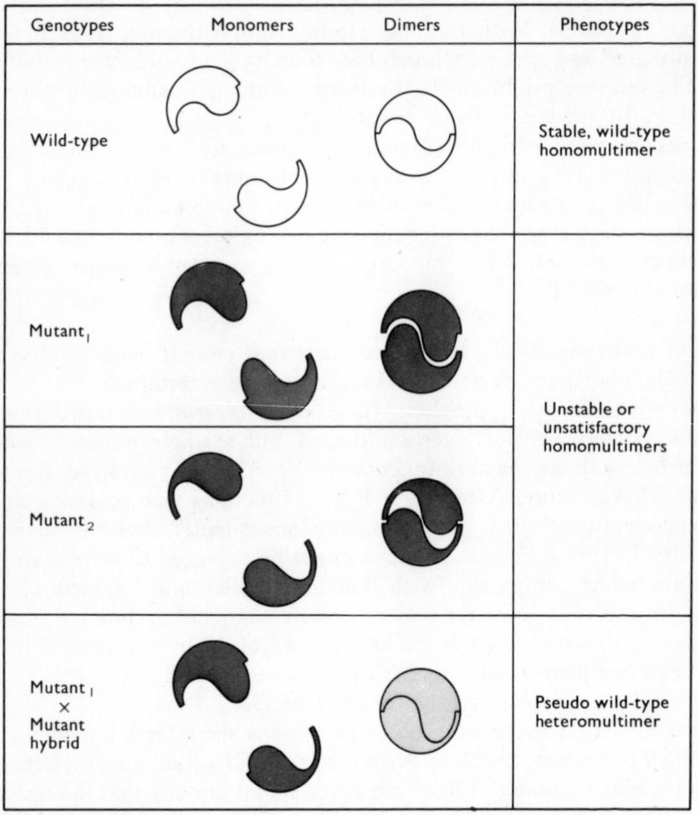

| Genotypes | Monomers | Dimers | Phenotypes |
|---|---|---|---|
| Wild-type | | | Stable, wild-type homomultimer |
| Mutant₁ | | | Unstable or unsatisfactory homomultimers |
| Mutant₂ | | | |
| Mutant₁ × Mutant hybrid | | | Pseudo wild-type heteromultimer |

**Fig. 8.4** Schematic representation of intracistronic complementation based on a homomultimeric quaternary structure for the wild-type enzyme.

two types of mutant homopolypeptide as well. And the relative stability of the three proteins may be an important element with regard to the level of wild-type activity.[81]

A considerable body of evidence is now accumulating in support of the 'hybrid protein' hypothesis. Indeed, in certain cases, complementation has been demonstrated *in vitro* following the mixing of cell-free extracts or purified proteins from complementing mutants, the resulting enzymes being comparable to those formed *in vivo*. Clearly, therefore, complementation can occur in the absence of protein synthesis.[20]

In summary, trans hybrids between recessive mutants which are defective in functionally non-allelic cistrons are expected to be wild-type on the basis of simple additive action, each mutant contributing a different normal enzyme or else a different normal component (polypeptide) of a heteromultimeric protein. Trans hybrids between mutants which are defective in allelic cistrons are not expected to be wild-type because neither contributes a normal enzyme or normal component. However, a small proportion of such hybrids will approach wild-type if their differently defective polypeptides interact favourably in the formation of a "heteromultimeric" complex.

## INTERACTIONS BETWEEN STRUCTURAL AND CONTROLLING GENES

The discussion so far has been concerned mainly with the control of the nature of the gene product. But while some genes may function all the time and all genes function some of the time, all genes do not function all the time. And the interactions considered here are concerned with the temporal aspects of gene activity.[83]

### Enzyme induction

In many micro-organisms, the inherent capacity to produce certain katabolic enzymes is only barely expressed in the absence of a highly specific inducer. This is usually the substrate of the pathway but related compounds which cannot be katabolized may be even more effective in this respect.[86]

For example, the β-galactosidase of *E. coli* catalyses the hydrolysis of lactose to glucose and galactose. While this enzyme is barely detectable in wild-type strains grown without β-galactosides, its activity increases by as much as ten thousand times in their presence, and falls quickly when they are removed (Fig. 8.5).

This induction is associated with two other remarkable properties:

1. Irrespective of the inducer employed, induction leads not only to an elevation of β-galactosidase activity but to the co-ordinate induction of two other enzymes, β-galactoside permease, which allows β-galactosides to enter and accumulate in the cells, and β-galactoside acetylase. Co-ordinate induction does not mean that these enzymes are produced in equimolecular amounts. But whatever the absolute level of induction, the

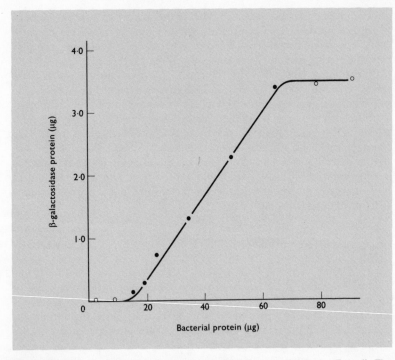

**Fig. 8.5** The kinetics of induced β-galactosidase synthesis in *E. coli*. The accumulation of enzyme is expressed as a function of increase in cell mass. Solid circles indicate the period during which the inducer was present. Over the ascending part of the curve the enzyme represents 6·6 % of the total protein.

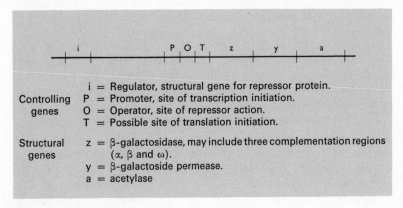

**Fig. 8.6** The lactose operon of *E. coli* and its regulator gene.

relative quantities of these enzymes remain the same. In the case of β-galactosidase and the permease this ratio is about 25:1 on a molar basis.

2. The structural genes which determine these enzymes form a contiguous group on the linkage map (Fig. 8.6).

| GENES | ENZYMES | ORDER OF ACTION |
|---|---|---|
| O | Operator | |
| G | Pyrophosphorylase | 1 |
| D | Dehydrogenase | 10, 11 |
| C | Transaminase | 8 |
| B | Dehydrase, phosphatase | 7, 9 |
| H | Amidotransferase | 5 |
| A | Isomerase | 4 |
| F | Cyclase | 6 |
| I | Hydrolase | 3 |
| E | Pyrophosphohydrolase | 2 |

**Fig. 8.7** The histidine operon of *Salmonella*. The two functions (7 and 9) shown by enzyme B are probably different activities of the same protein. The position of a gene in an operon and hence the location of its mRNA in the poly-cistronic product of the operon may affect the frequency of translation. This, in turn, will affect the molar ratios of the enzymes subjected to co-ordinate regulation.

## Enzyme repression

The opposite, yet comparable, condition is known for many anabolic enzyme systems. For example, all the enzymes involved in histidine synthesis in *Salmonella* are depressed in a co-ordinate manner by histidine, the end product of the pathway. This depression is quite distinct from end product (feedback) inhibition which operates by inhibiting the activity of the enzyme which acts first, or one which functions early in the pathway. Rather, end product repression, like precursor induction, operates at the level of enzyme synthesis and the structural genes concerned form a contiguous array (Fig. 8.7). Concatenated clusters of genes which are subject to co-ordinate induction or repression constitute an operon.[3]

Operon organizations are common in bacteria and less so in viruses.[31] But they have not been clearly shown in higher organisms (see p. 219).

## The operon hypothesis

Jacob and Monod proposed that co-ordinate inductions and repressions were based on the fact that the concatenated array of the structural genes concerned constituted a single unit of transcription.[57] On this view, each reading of the operon produced a single, polycistronic mRNA so that all the structural genes in the operon were transcribed to the same extent.[55]

In fact, it is now known that the mRNAs corresponding to the lactose and tryptophan operons of *E. coli* and the histidine operon of *S. typhimurium* are of a size appropriate for their entire respective operons. What is more, the dimensions of the polyribosomes which support β-galactosidase synthesis indicate that the polycistronic mRNA remains intact during translation. However, it cannot be regarded as a unit of translation in the sense that an operon is a unit of transcription because all the cistron equivalents in polycistronic mRNA are not translated to the same extent.

On the above hypothesis, regulation is effected at the level of transcription, induction involving a large and rapid increase in the rate of production of specific polycistronic mRNA while a corresponding decrease is implicated in repression. Studies on mutant strains have revealed the role of the effector molecules and the nature of the processes which effect a temporal control of gene activity in bacteria.[15]

Certain mutant strains of *E. coli* produce high levels of the enzymes of the *lac* operon even in the absence of inducers. These strains are of two main kinds, operator constitutive ($o^c$) and regulator constitutive ($i^c$). The mutations involved in the former map in a region to the left of, and adjacent to, the $z$ gene. Those of the latter also map to the left of $z$ but not in a contiguous region. These two classes of mutants differ not only in location but in their dominance relations, $o^c$ mutations being dominant while $i^c$ mutations are not. They differ also in regard to the phenotypes they determine in cis-trans hybrids produced on crossing with strains which are defective in structural genes of the operon (Fig. 8.8).

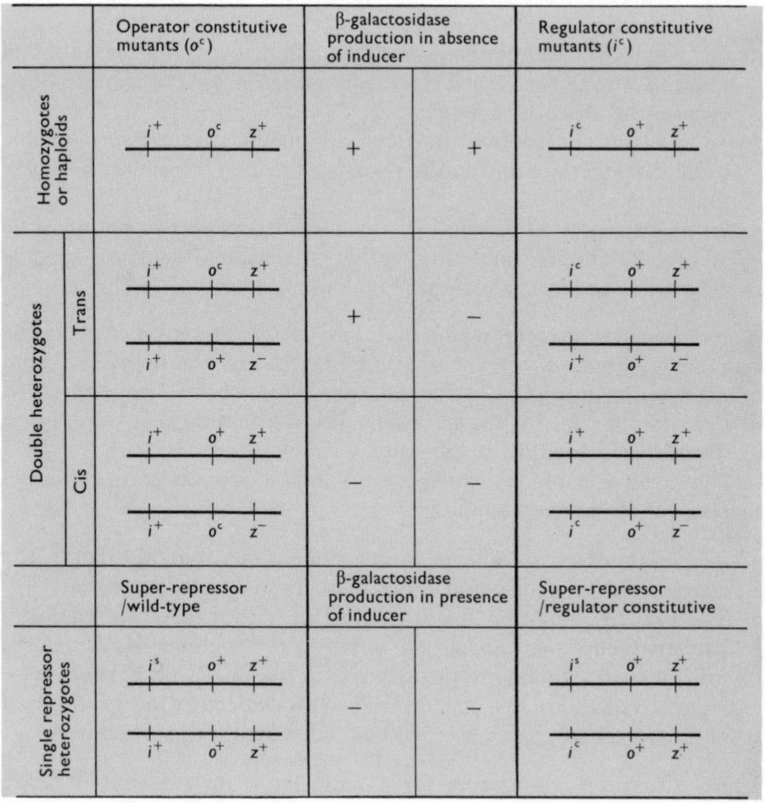

| | | Operator constitutive mutants ($o^c$) | β-galactosidase production in absence of inducer | | Regulator constitutive mutants ($i^c$) |
|---|---|---|---|---|---|
| Homozygotes or haploids | | $i^+$   $o^c$   $z^+$ | + | + | $i^c$   $o^+$   $z^+$ |
| Double heterozygotes | Trans | $i^+$   $o^c$   $z^+$ <br> $i^+$   $o^+$   $z^-$ | + | − | $i^c$   $o^+$   $z^+$ <br> $i^+$   $o^+$   $z^-$ |
| Double heterozygotes | Cis | $i^+$   $o^+$   $z^+$ <br> $i^+$   $o^c$   $z^-$ | − | − | $i^+$   $o^+$   $z^+$ <br> $i^c$   $o^+$   $z^-$ |
| | | Super-repressor /wild-type | β-galactosidase production in presence of inducer | | Super-repressor /regulator constitutive |
| Single repressor heterozygotes | | $i^s$   $o^+$   $z^+$ <br> $i^+$   $o^+$   $z^+$ | − | − | $i^s$   $o^+$   $z^+$ <br> $i^c$   $o^+$   $z^+$ |

**Fig. 8.8** The differences between operator and regulator constitutive mutants with regard to dominance and cis-trans position effects (cf. Figs. 8.6 and 8.9). Plus = Constitutive, Minus = Inducible (Wild-type).

Other variant strains are unable to produce the *lac* enzymes even when inducer is supplied. This super repressed state also is determined by mutation in the *i* gene ($i^s$) which proves to be dominant to both $i^+$ and $i^c$.

These and other related observations can be explained on the following hypothesis of Jacob and Monod:

1. In the wild-type, the regulator gene, $i^+$, produces a *lac* specific repressor which binds to the operator, $o^+$, thereby preventing the transcription of the operon.[70]
2. In the wild-type, the inducer molecule associates with the repressor and so prevents it from binding or leads to the release from operator of previously bound repressor.
3. In operator constitutive strains, $o^c$, the mutated operator region shows reduced affinity for normal repressor and so the operon is transcribed even in the absence of inducer.
4. In regulator constitutive strains, $i^c$, the modified repressor is unable to bind to operator and so the releasing function of inducer is superfluous.
5. In super-repressed strains, $i^s$, the modified repressor shows little affinity for inducer molecules and so it remains bound to operator even when inducer is present (Fig. 8.9).

The repressor, a protein, with a molecular weight of $1.5-2.0 \times 10^5$ in the case of *lac*, binds directly to the duplex DNA of the operator region which must be at least 12 base pairs long. Thus, while $i$ can be regarded as a controlling gene for the *lac* operon, it is a structural gene with regard to the primary structure of the repressor molecule.[35]

The rapid response shown by most inducible systems on the addition or removal of inducer shows that:

1. The number of specific repressor molecules cannot be large, the estimate for the *lac* operon being about 10 molecules per $i$ gene,
2. The mRNA cannot be long lived. In fact, most mRNA molecules in bacteria seem to have a half life of only a few minutes. However, the mRNA corresponding to penicillinase in *Bacillus cereus* appears to be longer lived and in this case the induced state persists for the duration of the division cycle even when new RNA synthesis is inhibited.

It has also been shown that DNA extracted from substituted phage particles carrying the *lac* operon in the $o^c$ state shows little affinity for *lac* repressor molecules obtained from uninduced wild-type strains. Further, isopropylthiogalactoside (IPTG), a non-substrate inducer of the *lac* operon, is known to cause the specific release of *lac* repressor from the *lac* region *in vitro*.

Being diffusible and cytoplasmic, the protein products of regulator genes can influence their subservient operons whatever their relative locations in the genome. Consequently, regulator mutations do not show cis-trans position effects. Operators, on the other hand, are contiguous with the structural complex they control and they occupy the end of the operon from which its transcription proceeds. Presumably, therefore,

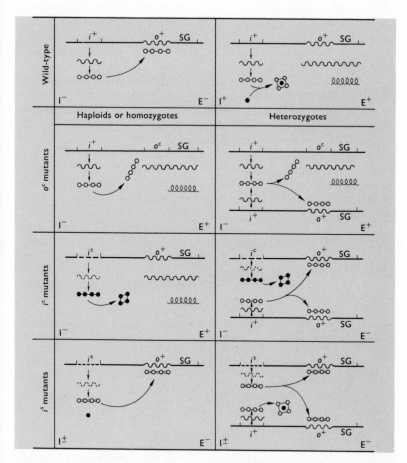

**Fig. 8.9** The operon hypothesis and the effects of operator and regulator mutations in relation to an inducible system (schematic). See text and compare Figs. 8.6 and 8.8.

operators function by virtue of their position near the point of transcription initiation and the cis-trans effect they show is a reflection of this.

However, while the operator is the site of action of repressor, the actual site of transcription initiation (promoter) probably lies slightly to the left in the *lac* operon and it may represent a binding site for RNA polymerase. On this view, the binding of repressor to operator prevents the movement of the polymerase along the operon. Mutations at the promoter site (*p*) affect the maximum rate of transcription.

End product repression of operon transcription can be explained on a parallel hypothesis. In this case, however, it is the repressor-end product complex which binds to the operator while unbound repressor does not. Consequently, the operon is transcribed only in the absence of end product.

The systems outlined above represent negative control in that the basic state of the operon is assumed to be 'on' and its regulation is effected by turning it 'off'. The opposite condition, positive control, is also known.[29] Thus, in the case of the L-arabinose operon ($BAD$) of *E. coli*, the repressor-inducer complex appears to function as a positive activator (see also p. 214). This serves to stimulate the production of a kinase ($B$), an isomerase ($A$) and an epimerase ($D$) which are specified by contiguous genes. The production of a permease ($E$) specified by an unlinked gene is also stimulated. In fact, proximity is not an indispensible condition for joint regulation.

For example, in both *E. coli* and *Salmonella*, four of the eight genes which specify enzymes involved in arginine synthesis are closely linked but the remainder occupy scattered locations. Nevertheless all these genes are repressed by arginine and mutation at another locus ($arg^R$) releases them all from end product control. It would appear, therefore,

| $\beta$-gal | Inducible enzyme | Genotypes | Temperate phage | Lambda phage |
|---|---|---|---|---|
| + | No enzyme in absence of effector (substrate) | $R^+$    $o^+$ SG | No infective particles without induction (lysogeny: prophage) | + |
| | Enzyme produced on induction | $R^+$    $o^+$ SG | Lysis and infective phage on induction (vegetative state) | |
| $o^c$ | Constitutive enzyme (dominant to $o^+$) | $R^+$    $o^c$ SG | Lysis and infective phage (virulent to lysogenic bacteria) | $c_1$ |
| $i^c$ | Constitutive enzyme (recessive to $R^+$ and $R^s$) | $R^c$    $o^+$ SG | Lysis and infective phage (not virulent to lysogenic bacteria) | $c^-$ |
| $i^s$ | Enzyme negative non-inducible (dominant to $R^+$ and $R^c$) | $R^s$    $o^+$ SG | Lysogenic non-inducible (dominant to $R^+$ and $R^c$) | ind$^-$ |

**Fig. 8.10** A comparison of *E. coli* and $\lambda$-phage mutants in relation to the operon model.

that arginine and the product of the wild-type $arg^R$ locus function as co-repressors to which a series of unlinked genes is subject, possibly because the separate structural genes each have an operator region identical with that of the others. However, while these scattered genes are turned on and off at the same time, their regulation is not co-ordinate because they are not all transcribed to the same extent. This lends further support for the view that the production of polycistronic mRNA is the basis of truly co-ordinate induction and repression.

Mutations affecting various properties of temperate phage particles are known which closely resemble those described above in relation to enzyme-controlling operons. Consequently, the regulation of such phage features as virulence and inducibility have been described in terms of an operon organization (Fig. 8.10).

## SUPPRESSOR MUTATIONS

Mutants sometimes revert to wild-type. This change may be due to exact back mutation in which case the revertant is identical with the original wild-type. However exact reversal of the original mutation need not be involved but even in this case the revertant may be distinguishable from the true wild-type only with difficulty. For example, reversions of missense mutations may exploit the degenerate nature of the genetic code as follows:

| Wild-type | UCU | Serine |
|---|---|---|
|  | ↓ |  |
| Mutant | UGU | Cysteine |
|  | ↓ |  |
| Revertant | AGU | Serine |

Further, we have already seen that plus-minus pairs of frame-shift mutations within a cistron can have mutually suppressing effects if they occur fairly close together. In this case, however, though the revertants may be superficially wild-type, they are actually pseudo-wilds (see p. 166).

But the suppressors we are concerned with here are those which occur outside the mutant genes they suppress. These too confer pseudo-wild phenotypes though the original mutant gene is unaltered. The suppressors which come into this category are themselves a heterogeneous assemblage in more than one respect. For example, some of them affect more than one functional region (super-suppressors)[47] while others are specific to the extent of suppressing only one or some mutant alleles of a particular gene (Table 8.2).

**Table 8.2** The specificity of suppressors of tryptophane synthetase deficiency (*tryp*-3 mutants) in *Neurospora*. Suppression is indicated by plus. Suppressor mutations $su_3$ and $su_{24}$ are linked, the other combinations, unlinked. (Data of Suskind, S. R. (1957) Gene function and enzyme formation, 123–9. In *A Symposium on the Chemical Basis of Heredity*, McElroy, W. D. and Glass, B., Johns Hopkins University Press, Baltimore.)

| Suppressor mutations | $td^2$ | *Tryp-3 mutations* $td^3$ | $td^6$ | $td^{24}$ |
|:---:|:---:|:---:|:---:|:---:|
| $su_2$ | + | − | − | − |
| $su_6$ | + | − | + | − |
| $su_3$ | − | + | − | + |
| $su_{24}$ | − | + | − | + |

The mode of action of suppressor mutations is various also.[27] Thus, some are known to act by opening up alternative pathways for the synthesis of end product while others release the offending enzyme from its susceptibility to inhibitors. Of particular interest, however, are the suppressors which influence the translation process itself and so actually alter the primary structure of the mutant polypeptide.

We have already seen that amber mutants of $T_4$ phage produce only amino fragments of head protein when they grow on restrictive ($su^-$) hosts (p. 173). Permissive or non-restrictive hosts ($su^+$), however, allow the synthesis of complete polypeptides. In these, the amber codon (UAG) is evidently translated so that it does not function as a stop signal.

A comparable situation occurs in regard to other amber mutations, for example, those in the alkaline phosphatase of *E. coli* or the coat protein of $f_2$ phage. But the really remarkable feature of amber suppression is this. In a given permissive host, the amino acid inserted at the amber codon is the same irrespective of the structural gene involved, the position of the amber codon in it, or the amino acid occupying that site in the wild-type. However, the inserted amino acid varies from one permissive host to another and the level of suppression can vary from 15–65%.

It would appear, therefore, that the amber codon is ambiguous in that it is differently translated in non-permissive and permissive hosts and the latter differ among themselves in this respect.[92]

Studies on the *in vitro* synthesis of coat protein in an amber mutant of $f_2$ phage have shown that the $su_1^+$ permissive host produces a variety of serine-accepting tRNA which is not present in non-permissive hosts. In the absence of this molecule, amino fragments are formed which terminate at the site of the amber mutation. But in its presence, serine is incorporated at this location.

Similarly, $su_{III}^-$ hosts produce a tyr-accepting tRNA which recognizes the amber codon UAG but does not appear to respond to the normal *tyr* codons UAU and UAC. In this case it is known that a change in the anticodon of the tRNA is responsible, the GUA of $su_{III}^-$tyr-tRNA being changed to CUA in the tyr-tRNA of $su_{III}^+$ (single base transversion). And CUA is the anticodon of the amber codon, UAG. (Nucleotide sequences are conventionally written in the $5'-3'$ direction and, because of the anti-parallel nature of pairing between nucleotide sequences, in the direction in which they are written, bases 1, 2 and 3 of the codon correspond with bases 3, 2 and 1 respectively of the anticodon.) Clearly, the permissive host must also produce other varieties of tyr-tRNA which can recognize the normal *tyr* codons, UAU and UAC, for otherwise the $su_{III}^+$ would be lethal.

A similar basis has been revealed for the suppression of missense mutations. Thus, a *gly* → *arg* mutation in the A protein of tryptophan synthetase is suppressed in $su_{36}^+$ hosts which produce an ambiguous gly-tRNA capable of recognizing the *arg* codon AGA (see p. 171).

## POSITION EFFECTS

We have already considered some situations in which the effects of a mutation depend not only on its own nature and location but on its position relative to other genetic states in the genome (Table 8.3).

Where two or more mutant sites in different structural genes are concerned, a wild phenotype is expected whenever these sites occur in the wild-type state. In other words, the phenotype can be predicted from a genotypic formula which makes no reference to the spatial relations of the mutational states.[63]

However, exceptions to this general rule are known in cellular organisms. Some of these can, perhaps, be accommodated by redefining the limits of the structural genes concerned. But in others a definite position effect must be accepted.[40]

### *Drosophila*

A tandem duplication of a short segment of the X chromosome in *Drosophila* leads to a reduction in the number of eye facets (Bar eye). Unequal crossing-over in females homozygous for the duplication can lead to a triplication of the segment concerned. The segment shows a dose effect: the more often it is represented, the greater the reduction in the number of facets. But although the segment is represented equally in

**Table 8.3** A summary of position effects (see also Figs. 8.11–8.15).

| Relative locations of mutations / Types of mutations | Same operon | | | | Separate cistrons | Separate operons |
|---|---|---|---|---|---|---|
| | Same cistron | | Nonsense | Missense | $o^c$-SG | General |
| | Frame-Shift Addition/Deletion | | | | | |
| | Proximal | Distal | | | | |
| **Haploids or Homozygotes** | | | | | | |
| $\dfrac{+}{+}$ | + | + | + | + | + (Inducible E) | + |
| $\dfrac{m_1}{+}$ | − | − | − | − | − (Constitutive E) | − |
| $\dfrac{+}{m_2}$ | − | − | − | − | − (Inducible CRM) | − |
| $\dfrac{m_1}{m_2}$ | (+) | − | − | − | − (Constitutive CRM) | − |
| **Heterozygotes** Trans | | | | | | |
| $\dfrac{m_1 \quad +}{+ \quad m_2}$ | − | − | − | − or (+) | − (Constitutive E, Inducible CRM) | + |
| Cis | | | | | | |
| $\dfrac{m_1 \quad m_2}{+ \quad +}$ | + | + | + | + | + (Inducible E, Constitutive CRM) | + |

Bar homozygotes and triple/single heterozygotes, the extent of the eye reduction is significantly greater in the latter (Fig. 8.11). This effect is called a stable position effect (cf. V-type position effects).

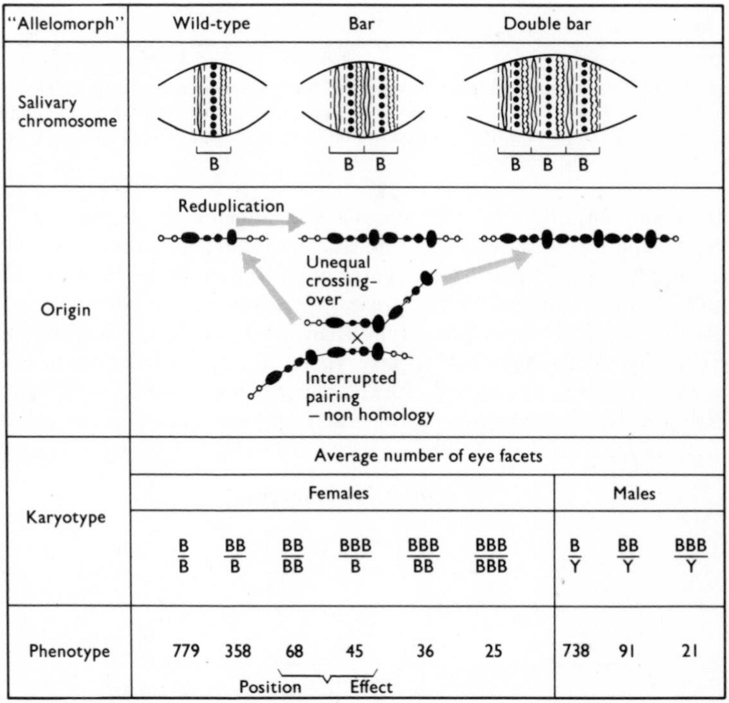

| "Allelomorph" | Wild-type | Bar | Double bar |
|---|---|---|---|
| Salivary chromosome | B | B  B | B  B  B |

(Origin)

Reduplication
Unequal crossing-over
Interrupted pairing – non homology

| Karyotype | Average number of eye facets | | | | | | | | |
|---|---|---|---|---|---|---|---|---|---|
| | Females | | | | | | Males | | |
| | $\frac{B}{B}$ | $\frac{BB}{B}$ | $\frac{BB}{BB}$ | $\frac{BBB}{B}$ | $\frac{BBB}{BB}$ | $\frac{BBB}{BBB}$ | $\frac{B}{Y}$ | $\frac{BB}{Y}$ | $\frac{BBB}{Y}$ |
| Phenotype | 779 | 358 | 68 | 45 | 36 | 25 | 738 | 91 | 21 |

Position  Effect

Fig. 8.11 The cytological basis of the Bar mutations in *Drosophila*. (From McLean, R. C. and Ivimey-Cook, W. R. (1967), *Textbook of Theoretical Botany*, Vol. III, Longmans Green, London.)

Position effects of a different kind have been described in connection with bithorax mutations in *Drosophila*. Mutations in this series form a closely linked cluster on chromosome 3 and they all have related and manifold effects on the development of the meso- and meta-thoracic segments (Fig. 8.12). Their interest in the present connection is that while some pairs of mutations (*bx* with either *pbx* or *bxd*) give wild phenotypes in both the cis and trans configurations, others (e.g. *bxd* with *pbx*) give this phenotype only in cis dihybrids.

To this extent the results are similar to those in regard to complementing and non-complementing intracistronic mutations. But there is a

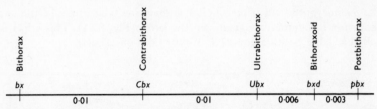

**Fig. 8.12**  The bithorax region at locus 58·8 of Chromosome III in *Drosophila melanogaster*. Mutations in this region have various manifold effects on the wings, halteres, thorax, legs and the first abdominal segment. Flies homozygous for both *bx* and *pbx* are four-winged.

further anomaly. By gross chromosome rearrangement, a segment which includes the bithorax region can be translocated to a distant location in the complement. Trans heterozygotes of the constitution $bx +/+ Ubx$ which are homozygous for such rearrangements do not show a greater departure from the wild phenotype than those which are homozygous for the original arrangement. Clearly, therefore, the rearrangement itself is without effect in this regard. But the phenotypes of trans heterozygotes which are also heterozygous for the rearrangement show an even greater departure from wild-type. Thus

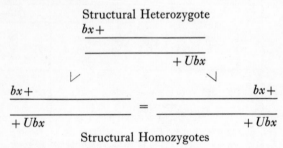

Thus, the relative location in the complement of the two differently-defective bithorax regions is important in relation to their effects.

In *Drosophila*, homologous chromosomes are often closely paired in parallel even in somatic cells (see p. 203) and structural heterozygosity is expected to interfere with the incidence and intimacy of this somatic pairing. Consequently, there is reason to believe that the $bx +$ and $+ Ubx$ segments are likely to be much closer to each other in the metabolically active nuclei of structural homozygotes than of structural heterozygotes. And proximity in the active nucleus may be the basis of both types of position effect.

Yet other situations are known in *Drosophila* where rearrangements *per se* have an effect. For example, in the homozygous condition, the

X-linked mutation $w$ gives a white eye but red wild-type eyes are found in $+^w$ $w$ heterozygotes. However, when the wild-type gene is translocated to a position near procentric heterochromatin, its pigment-producing capacity is impaired. But this inhibition is variable as between cells even within the individual. Consequently, while some eye facets do produce the non-diffusible pigment, others do not and the result is a variegated eye. This variegated (V-type) position effect is shown by both structural heterozygotes and those homozygous for the rearrangement. In this case, therefore, a V-type position effect is superimposed on a cis-trans position effect because the phenotypes of $w/+^w$ flies depend on whether $w$ or $+^w$ is coupled with the rearrangement (Fig. 8.13).

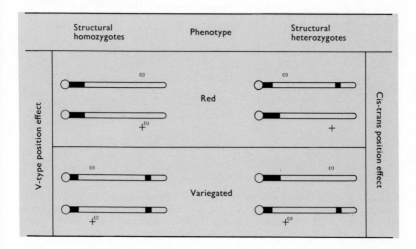

**Fig. 8.13** V-type and cis-trans position effects following the transposition of genes to an unaccustomed position near procentric heterochromatin. The effect is illustrated with reference to the white eye gene on the X chromosome of *Drosophila melanogaster.*

Variegation of this kind is associated with a spreading effect in that:

1. The incidence and extent of inhibition decrease with increasing distance from the heterochromatin, and
2. In a given cell, one gene cannot be inhibited unless and until the other genes which lie between it and the heterochromatin are affected. But genes beyond it may or may not be affected according to the extent of the spreading effect (Fig. 8.14).

When cells showing phenotypic anomalies cannot be used for sexual reproduction, it is not easy to distinguish between altered genes and genes of altered action or interaction. Thus, it could be argued that the proximity of heterochromatin in the above example causes gene mutation in somatic cells. What is clear, however, is that when the offending gene is removed from the proximity of heterochromatin (by crossing-over or further rearrangement), normal function is restored.

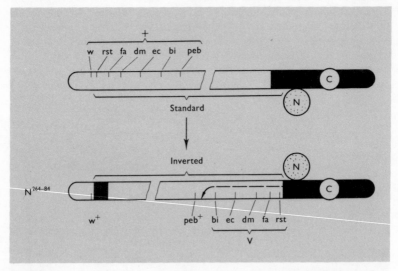

**Fig. 8.14** The spreading effect in relation to V-type position effects in *Drosophila*. The positions of the centromere (C) and the nucleolus (N) of the X chromosome are indicated and the heterochromatic regions are shown solid. (From Lewis, K. R. and John, B. (1963), *Chromosome Marker*, J. and A. Churchill, London.)

Similar results are sometimes obtained when genes normally located near heterochromatin (e.g. *light* and *peach*) are translocated away from it. V-type position effects have been described in maize and the evening primrose and certain of its properties are found also in mammalian females.

## Female mammals and male mealybugs

The presence of a sex chromatin (Barr) body in somatic interphase nuclei of female mammals was first reported in 1949. It was not until 1959, however, that Ohno and his colleagues showed that this body represented an entire heterochromatic X chromosome. Sex chromatin is not present in either male or female early embryos recovered from the fallopian tubes of mice, rats or hamsters. All implanted embryos with an XX constitution, however, show a single condensed X in every cell. The female zygote thus begins its development with both X chromosomes fully extended during interphase. But at about the time of implantation one or other of the two X chromosomes in every cell becomes positively heteropycnotic. Once determined this behaviour is fixed and irreversible so far as somatic cells are concerned. In the female germ cells, however, the affected X appears to revert to an isopycnotic state.

Numerous investigations on both normal and subnormal sex genotypes have shown that the number of Barr bodies follows the rule

$$B = n - \frac{p}{2}$$

where n = number of X chromosomes per cell, and p = the level of ploidy of that cell. Coupled with this behaviour it can be shown autoradiographically that the condensed X does not synthesize RNA so that the two X chromosomes of a normal female are distinguishable both cytologically and cytochemically. Abnormal XO karyotypes develop into females which do not show Barr bodies. Moreover, in mouse, though not in man, these females are normal and fertile. Indeed in the creeping vole (*Microtus oregoni*) the female soma is regularly XO in constitution from the beginning.

These findings have been used in support of the hypothesis that one of the two X chromosomes of female mammals is epigenetically inactivated during early development and this X forms the sex chromatin body.[65] This explanation can be extended to accommodate the fact that, in mouse, the X-linked coat colour genes *tabby* and *mottled* regularly produce a dappled or variegated pattern in heterozygous females. Very few autosomal coat colour genes, on the other hand, give mottled effects and these are shown equally in both sexes. To explain this situation one needs to assume that the inactive X is selected on a random basis and that the inactivation is both stable and transmissible in cell heredity. These in turn imply that, for certain genes at least, the female mammal is a natural genetic mosaic. In the case of genes with non-diffusible products, therefore, both variegation and cis-trans position effects can be expected in chromosomally normal females (Fig. 8.15).

When, following chromosome breakage, autosomal genes are translocated to the X chromosome they tend to be subjected to the same hemizygous inactivation. However, it would appear that the incidence and intensity of this inhibition is a function of the distance between the locus and point of breakage. This effect of intercept length is comparable with the gradient observed in relation to V-type position effects in *Drosophila*. It has led to the suggestion that inactivation spreads along the X from control centres whose effects are chromosome limited. In the

rearrangements studied to date, centric autosomal segments are not inactivated when X material is translocated to them. Further, although
maternal and paternal chromosomes appear to be inactivated at random
in normal XX females, preferential inactivation is found in females with
certain X-autosome rearrangements. Alternatively, the inequality in the
numbers of differently inactivated cells could be due to their differential
survival and division following random inactivation. This too can lead

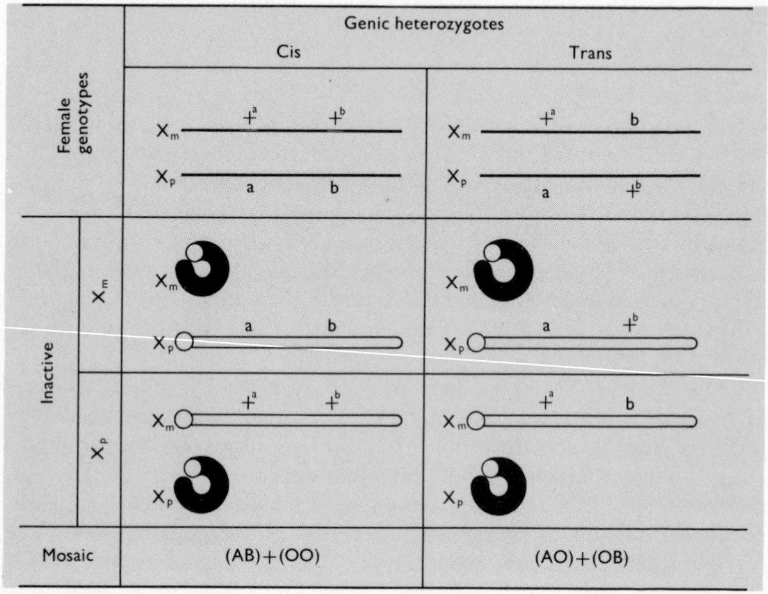

Fig. 8.15 The mosaicism and cis-trans position effects in respect of genes with
non-diffusible products determined by random, hemizygous X inactivation in
female mammals. Paternal inactivation (last row) applies to all chromosomes in
male mealybugs where cis-trans effects are expected without mosaicism.

to cis-trans position effects. It will be appreciated that the extended state
of an X does not mean that all its genes are active all the time. Its
extension simply means that, unlike the pycnotic X, it is amenable to
locus by locus control. Indeed, there is an opinion that even the condensed X shows some activity and, significantly, it appears to contribute
to nucleolar organization in some cells of the female mouse.

However one interprets the data in detail the available information
points to a block regulation mechanism. A comparable system operates

in male mealybugs with a lecanoid chromosome system. Here one entire haploid set of chromosomes, the paternally derived one, becomes positively heteropycnotic early in male embryogeny whereas in females both sets remain euchromatic. This H-set is incapable of synthesizing RNA and appears to be developmentally inert. Indeed at male meiosis the H-set segregates from the euchromatic set and only those meiotic products with the E-set form sperm. This E-set will of course become the H-set in the next generation of males. In a number of tissues, including some cells of the skeletal muscles, the intestinal tract and all the cells of the Malphighian tubules and the cyst wall cells of the testes, the heterochromatinization of the paternal set is reversed and this happens at different times in different tissues. Reversal often occurs in diploid cells which later undergo endopolyploidization.

Since heterochromatinization affects a particular chromosome set, it does not lead to mosaicism. But the phenotypes of the males, like those of female mammals in certain cases, vary according to the origin, maternal or paternal, of the genes concerned. In other words, a 'cis-trans' effect can be expected even for genes on different chromosomes! Further, the males breed and develop as though they had been derived by haploid parthenogenesis.


## PUFFING IN POLYTENE CHROMOSOMES

Control mechanisms in animals can affect groups of genes which act in sequence. This is evident from gene activity in cells which are already determined with regard to their differentiation. The clearest examples of this are found in the control of puffing in polytene chromosomes.

The usual criterion for deciding whether a particular gene is active or not is to determine whether the final product of that gene is present or absent. Polytene systems, however, provide an opportunity for studying primary gene action. In the normal cell cycle, interphase DNA replication and chromosome duplication are followed by the individualization of the products of these processes and their subsequent separation during nuclear division. But this sequence is not obligatory. The products may remain associated and continue to duplicate in the absence of nuclear division. The net result is the production of a polytene or multistranded chromosome in which many hundreds or even thousands of identical threads are arranged side by side. This, in effect, magnifies the detailed structure of the chromosome by 2–3 orders of magnitude for the chromosomes increase both in width and length. This process of polytenization is best known in the larval development of dipteran flies where it is accompanied by an intimate association of homologous chromosomes so that each polytene unit is composed of a pair of multistranded homologous chromosomes. In these flies the polytene state is characterized also by transverse banding owing to the alternation of chromatic (band) and achromatic (interband) zones (Fig. 8.16). Moreover, these giant chromosomes remain visible as discrete entities throughout a large part of the life cycle. The bands differ so much in form and sequence that even very short chromosome segments have a distinctive appearance. Further, the bands are so numerous that the patterns of their distribution constitute morphological maps of high resolution. Indeed, a single mutation can be allocated to a change in a particular band. In no case has it been necessary to allocate more than one Mendelian gene to any one band and there has been a tendency to assume an absolute equivalence of bands and genes. But the heterochromatic segments of the normal diploid set do not take part in the polytenization process.

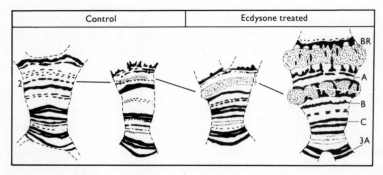

**Fig. 8.16**  The behaviour of puff locus IV-2-B in last instar larvae of *Chironomus tentans*. (Based on Clever, U. (1961). From Lewis, K. R. and John, B. (1963), *Chromosome Marker*, J. and A. Churchill, London.)

Although all the bands are represented in the polytene chromosomes of different tissues, they are not present in the same state. Indeed a given band may differ in appearance at different times in the same tissue. In both cases the differences involve the dispersion of the lateral components of the band so that the region becomes swollen. Such diffuse swellings are referred to as puffs. In chironomids some of the puffs are very large and are known as Balbiani rings. All the available evidence indicates that puffing is associated with increased activity. Thus, both histochemical and autoradiographic analyses give clear evidence for an augmentation of synthetic activity. Puffs are active in RNA synthesis, probably mRNA synthesis. They also accumulate proteins but these are not synthesized at the site of the puff and are probably of nucleolar origin.

Thus the evidence indicates that a locus is inactive when it occurs as a discrete band but active when that band is puffed. Inactive genes are thus highly condensed and genic activity is correlated with an extension of the material of the chromosome.

A few puffs are known to be involved in the production of specific products (Table 8.4) which evidently contribute to the properties of the secretion of the salivary gland. The intense puffing associated with moulting suggests that puffs might be concerned with the production of materials required for the histogenesis or histolysis which occur during pupation.

**Table 8.4**  Polytene puffs leading to the formation of specific products in salivary gland systems.

| Species | Puff | Product |
|---|---|---|
| *Chironomus pallidivitatus* | SZ | SZ granule |
| *Drosophila hydei* | 47B | Large mucopolysaccharide granules |
| *Acricotopus lucidus* | BR2 | Hydroxyproline in main and side lobes |

On Beermann's view a band represents a serial repetition of similar cistrons only one of which, the terminal or P segment, is transcribed into RNA. The remainder serve to bind the RNA produced by the P segment, presumably by DNA/RNA hybridization. This, in turn, holds the template RNA to the chromosome while it is complexed with protein for transportation to the site of translation in the cytoplasm. Perhaps, this packaging at the chromosome surface lies behind the position effects discussed earlier in connection with bithorax mutations. However, Ashburner has found that at least one puff in *D. melanogaster* (locus 7LCE in chromosome III) is divisible by translocation into two separate but co-ordinate puffs. In other words, a band and the corresponding puff, which on morphological and developmental criteria behave as a single unit, can be divided into 2 sub-units each of which organizes its own puff. Centromeres and nucleolar organizers are known to be similarly composed of supplementary units and, hence, to be divisible by breakage into fully functional sub-units.

It is of particular significance that puffing at particular bands depends on the prior activity of other regions. Further, the pattern of puffing can be influenced by hormone treatment. For example, when young larvae of *Drosophila* are treated with the moulting hormone, ecdysone, a puff

pattern similar to that which precedes normal pupation is prematurely
precipitated.


## THE AMPLIFICATION OF GENETIC INFORMATION

We have already seen evidence for the differential release of information during the process of development so that in different cells the
same genes differ in transcriptive ability even though they have the same
replicative functions. We turn now to consider cases where there is a
selective augmentation of information at specific sites based on differential replication.

A major event during oogenesis in animals is the synthesis of a large
reservoir of rRNA and ribosomes which is conserved to function during
early embryonic development. For example in the amphibian *Xenopus
laevis* new ribosomes are not formed in significant numbers until hatching
and mutant anucleolate embryos develop normally until the tail bud
stage. Protein synthesis in early embryogenesis is thus accomplished on
ribosomes formed during oogenesis and stored for later use. In at least
some cases it can be shown that this process is based on an augmentation
of those DNA templates which code specifically for the production of
nucleoli.[74] For example, in plethodont salamanders it involves the production and detachment of small circular segments of DNA which form
the basis for a large number of small nucleoli. This system of highly
selective gene replication represents a device for amplifying those DNA
units which code specifically for the rRNA required during early
development. Multiplication of the nucleolar-organizing region is known
to occur also during oogenesis in tipulid flies, beetles and crickets where
it leads to the production of so-called DNA bodies.

Differential epigenetic amplification of certain regions could, of course,
be superimposed on any duplication of loci which already exists in the
germ line complement. We have already seen, for example, that the basic
nucleolar organizer is itself composed of a number of supplementary
units and so it can be broken into two still functional and sufficient parts.

Further, in polyploids, all the chromosomes are represented three or
more times in the diplophase while in polysomics particular chromosomes
are disproportionately represented. In the case of duplications also dose
differentials are evident in the germ line as is the case with Bar Eye
(p. 195).

However, certain observations introduce the possibility of a different
kind of duplication of genetic information. For example, the type sub-
species of *Chironomus thummi* has 27% more DNA in the germline than
the more primitive *Ch. thummi piger* which has the same general karyo-

type. The DNA contents of salivary gland chromosomes with the same
level of polynemy show the same ratio of DNA values in the two sub-
species. The difference does not depend on a difference in band number
but on the fact the DNA content of certain bands in *thummi* exceed those
of their *piger* counterparts by some power of 2 up to $2^4$. This type of
duplication is, therefore, quite different from that in the case of Bar.[59]
Thus, if we regard the chromomere and the band to which it gives rise
as a unit of transcription, duplications of the Bar type increase the
number of these units while the *thummi* duplications increase their size.

In many animals, the chromosomes at diplotene, in particular during
meiosis in females, have a very characteristic 'lampbrush' appearance.
In newts, these chromosomes are very long (up to 800 μ) and distinctly
chromomeric. Paired lateral loops 15 μ or more in length arise from the
paired chromomeres of sister chromatids (Fig. 8.17). RNA synthesis

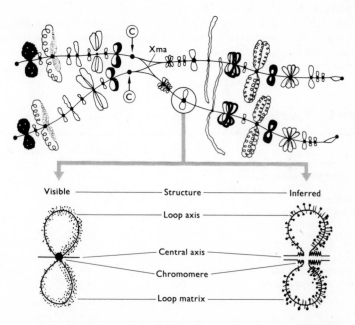

**Fig. 8.17** Diagrammatic representation of a lampbrush chromosome
of *Triturus*. This is a composite figure based on the work of H. G. Callan
and L. Lloyd. The lateral loops are of several different types by which
even short segments can be distinguished. Matrical material is laid
down on the loop axis in various ways but it is always asymmetrically
disposed. (From Lewis, K. R. and John, B. (1963), *Chromosome Marker*,
J. and A. Churchill, London.)

takes place on the DNA of the loop axis and labelled protein accumulates there also.[17]

On this and other evidence the loops have been regarded as the functional equivalents of the puffed bands of polytene chromosomes. In fact, the pattern of inheritance of loop heteromorphism follows that expected of Mendelian heterozygotes.

However, equating each chromomere, in band or loop, with a single gene introduces a problem because the length of the axial DNA in an average loop equals or exceeds that of the whole genoneme of *E. coli*. Such a length is sufficient to encode the information required to specify the products of some 3000 or more different structural genes.

We have already drawn attention to Beermann's proposal for resolving this difficulty (p. 205). But Callan has offered an alternative and, in some respects, opposite suggestion. Thus, he proposes that the loop includes a master gene and a whole series of identical slave genes which alone are responsible for RNA synthesis.

Whether extended loci of this kind are common or not cannot easily be assessed at the present time but they may explain the fact that even related species with similar chromosome complements can differ widely in DNA content. It may explain also why a lily or a lung fish has thirty times as much DNA as a man (Table 8.5).

**Table 8.5**  Variations in the DNA values of related species.

| | $2n = 2x$ | DNA ratio |
|---|---|---|
| Intra-specific : | | |
| *Chironomus thummi piger* | 8⎫ | |
| *thummi* | 8⎭ | 1 : 1·27 |
| Intra-generic: | | |
| *Rana temporaria* | 24⎫ | |
| *esculenta* | 24⎭ | 1 : 1·54 |
| *Vicia sativa* | 12⎫ | |
| *faba* | 12⎭ | 1 : 5·3 |
| Intra-familial (Ranunculaceae) : | | |
| *Aquilegia caerulea* × *chrysantha* | 14⎫ | |
| *Anemone tetrasepala* | 14⎭ | 1 : 42·3 |

## PARAMUTATION

Allelic genes may interact via the products of their activity in heterozygotes but they themselves are not expected to change as a result of this interaction. In certain special cases, however, changes do occur. The

clearest example of this is paramutation in *Zea mays*. For example, the
$R$ locus, which is responsible for the formation of anthocyanin pigment
in the aleurone layer of the grain and certain vegetative parts of the plant,
consists of a series of multiple, functional alleles which have various
effects on the intensity and distribution of pigment (Table 8.6).

**Table 8.6** The phenotypic effects of some *R* alleles in *Zea mays* (From Brink,
R. A. (1964) 23rd *Symposium of the Society for the Study of Development and
Growth*, 183–230.) ( +) = weakly paramutagenic.

| Allele | Phenotype | | | Para-mutable | Para-mutagenic |
|---|---|---|---|---|---|
| | Grain | Seedling | Anther | | |
| $R^r$ | Self-coloured in 2 or 3 doses, darkly mottled in 1 dose | Red | Red | + | − |
| $R^g$ | Self-coloured in 1, 2, or 3 doses | Green | Green | − | + or − |
| $R^{st}$ | Stippled | Green | Green | − | + |
| $R^{mb}$ | Marbled | Green | Green | − | + |
| $r^r$ | Colourless | Red | Red | − | − |
| $r^g$ | Colourless | Green | Green | − | − |
| Paramutant forms | | | | | |
| $R^{r'}$ | Self-coloured in 2 or 3 doses, lightly mottled in 1 dose | Red | Red | + | ( +) |
| $R^{g'}$ | Self-coloured in 2 or 3 doses, lightly mottled in 1 dose | Green | Green | + | ( +) |

When an $R^r$ allele is recovered from either $R^{st}\ R^r$ or $R^{mb}\ R^r$ heterozygotes, its pigment producing capacity in the next and subsequent
generations is reduced even in the absence of $R^{st}$ or $R^{mb}$ which are themselves not altered by association with $R^r$. $R^r$ is then said to have undergone
paramutation and the paramutated $R^r$ allele is designated as $R^{r'}$ or more

specifically $R^{r/st}$ or $R^{r/mb}$. Further, the $R^{r'}$ paramutant is itself weakly paramutagenic. Similar behaviour has been observed at the $B$ locus in maize which also controls anthocyanin production.

In the case of $R$, the paramutation process appears to occur in somatic cells while in the $B$ locus it probably takes place at meiosis. In neither case, however, has the mechanism been clarified.

## NUCLEAR-CYTOPLASMIC INTERACTIONS

The cytoplasm contains genetic elements the transmission of which is not usually controlled as rigorously as that of nuclear genes. The lack of regular segregation requires that each cytoplasmic gene be represented many times (p. 75). The cytoplasm is also the environment of the nucleus. Therefore, it affects the behaviour of the principal carrier of genetic information and is, in turn, affected by it. In considering the cytoplasm and its interactions with the nucleus, therefore, it is convenient to distinguish its genetic aspects from its role as an internal environment.[88]

Elements with genetic continuity and permanence and capable of mutation are found in the cytoplasm as well as in the nucleus. Those in the nucleus are unquestionably more numerous, more amenable to analysis and have received more attention. But genetic particles of at least 3 kinds are known in the cytoplasm. Two of these, chloroplasts and mitochondria, share an organization based on membranes and a series of functions concerned with electron transport, phosphorylation and the production of energy. The third, the centriolar apparatus, forms fibres and is present in all organisms which have ciliated or flagellated stages at some point in their life cycle.

All three categories of particle are self-replicating, self-perpetuating mutable units and they all contain DNA (Table 8.7). For example, the chloroplast of *Chlamydomonas* contains about 3% of the total cell DNA. Likewise, the mitochondrial system of mouse fibroblast L-cells collectively harbours some 0·2% of the total DNA of the cell. Both types of DNA differ in base composition and metabolic properties from corresponding nuclear DNA and mitochondrial DNA molecules form closed rings. This DNA is double-stranded and replicates semi-conservatively. Chloroplasts and mitochondria also contain RNA in the form of ribosome-like particles which presumably play a role in inherent protein synthesis.

As yet, however, the function of this organellar DNA has not been established. Since these organelles are self-replicating the DNA must play the key role in this respect. Indeed it has been suggested that whereas the replication of the chromosomes is geared to the cell division cycle, the presence of DNA in cytoplasmic organelles allows them to replicate in response to stimuli, or else at times, other than those that trigger chromosome replication. This affords a flexibility for organellar growth which is independent of the cell division cycle.

**Table 8.7**   DNA content of nucleoids. $1 \times 10^{-16}$ g of DNA is equivalent to about 100 structural genes. (From Granik, S. and Gibor, A. (1967), *Progress in Nucleic Acid Research and Molecular Biology*, **6**, 143–86.)

| Nucleoid | DNA per nucleoid ($10^{-16}$ g) |
|---|---|
| Mitochondria | |
| *Bos* (heart) | 0·5 |
| *Neurospora* | 0·2–1·8 |
| *Tetrahymena* | 3·7 |
| *Brassica* | |
| Chloroplasts | 5·0 |
| *Acetabularia* | 1–10 |
| *Nicotiana* | 80 |
| *Euglena* | 110 |
| *Vicia* | 150 |
| Centriole | |
| *Tetrahymena* | 2 |

Since ribosome-like units occur in both chloroplasts and mitochondria, they presumably have the essential mechanisms for both transcription and translation. In fact it is known that isolated chloroplasts of *Chlamydomonas* incorporate amino acids into polypeptides. Nevertheless both organelles do come under the control of the nucleus (see below). The nature of this system of nucleo-cytoplasmic interaction is as yet undefined. Possibly, the organellar DNA contains the structural genes necessary to produce the proteins required for organellar growth while the regulatory genes which call them into or out of action are borne in the chromosomes. Alternatively the nuclear genes may play a more direct role in synthesizing the mRNA molecules, or even the protein molecules, required for the growth of both categories of cytoplasmic organelles.

As far as mitochondria are concerned there is sound evidence that they contain DNA but no good evidence that this has any sequence homology with nuclear DNA. It is established that isolated mitochondria can synthesize RNA and that mitochondrial RNA has sequence homology with mitochondrial DNA. There is, however, little direct evidence as yet that this RNA plays an informational role in the synthesis of mito-chondrial proteins. Some mitochondrial proteins are certainly coded for by nuclear genes and are synthesized outside the mitochondria. But the nature of the protein synthesis which occurs inside the mitochondria is not known.

## Chromosomal and extrachromosomal genes

We have already seen that mutations in nuclear and cytoplasmic genes can determine very similar phenotypes (e.g. mitochondrial and plastid abnormalities, resistance to streptomycin etc.) which suggests that wild-type nuclear and cytoplasmic genes co-operate in producing particular traits.

Indeed, in some cases, a nuclear gene can suppress the mutant pheno-type determined by an abnormal extrachromosomal element. For example, the *poky* mitochondrial mutants of *Neurospora* are specifically suppressed by the nuclear gene *f* which is not known to affect any other respiratory mutants. Spectroscopic studies show that the cytochrome abnormalities are still present in *f*-suppressed *poky* mutants though their peripheral phenotype is wild-type. Clearly, therefore, the nuclear gene does not actually modify the action of the cytoplasmic gene in this case although it compensates for this action.

Not only the activity but the replication and mutation of cytoplasmic genes also can be affected by the nuclear genotype. Thus, we have seen that the killer trait in *Paramoecium* requires the presence of kappa particles (p. 76). But the retention and replication of kappa depend on a particular nuclear constitution. Thus, nuclear genes $K$, $S_1$ and $S_2$ must be present and it seems likely that certain other chromosomal genes are involved as well.

Further, mutant genes are known in maize, rice and barley which affect the phenotype of the chloroplasts and give variegated plants. The plastid changes concerned are stable and irreversible and, once induced, show extrachromosomal transmission even after the removal of the mutation-inducing nuclear gene.

## The cytoplasm as the nuclear environment

In bacteria, at least some of the mechanisms which regulate the syn-thesis of protein operate directly on the genes governing the transcription of DNA into RNA. Observations on plant and animal cells, however, indicate that protein synthesis can be regulated at the level of trans-lation.[19] These observations are of two main kinds:

### Enucleation studies

The alga *Acetabularia* though unicellular is between 3 and 5 cm long when mature and consists of a branched basal rhizoid and an elongate stalk which is terminated by a reproductive structure, the cap, whose morphology is genetically determined and species-specific. The formation of the cap is a complex morphogenetic operation involving the synthesis of specific enzymes and cap polysaccharides. These syntheses are precisely regulated.[44]

*Acetabularia* is also uninucleate. The nucleus lies in the tip of one of the rhizoids and the cell can thus be enucleated simply by cutting off the rhizoidal branch which contains the nucleus. However, even many weeks after enucleation, a perfectly normal, species-specific, cap can be still formed *de novo*. Clearly all the information necessary for the production of the cap must be transcribed and passed from the nucleus to the cytoplasm long before it is actually used. In turn it follows that the expression of the information required for the synthesis of the cap must be under the control of cytoplasmic regulatory mechanisms which can initiate, regulate and eventually suppress the synthesis of proteins on pre-existing RNA templates. This temporal control system operates, therefore, at the level of translation not transcription.[14]

What is true for *Acetabularia* appears also to hold for *Spirogyra* and *Stentor* both of which show normal physiological functions after enucleation. Likewise enucleated eggs of *Arbacia* are capable, after suitable stimulation, of forming a blastula simply by regulating the translation of information stored in the egg cytoplasm at the time of its formation (see p. 206).

In plant and animal cells, therefore, the time at which a particular gene is transcribed into RNA has no immediate connection with the time at which that RNA is translated into protein. Consequently, the RNA templates for protein synthesis must persist in the cytoplasm for long periods of time and the regulatory mechanisms which govern translation must be in the cytoplasm. Indeed there is a growing body of evidence that mRNA must be protected against intranuclear degradation before it can be transported to the cytoplasm. While bacterial studies suggest that transcription and translation are closely coupled, in plants and animals they clearly need not be. In both *Acetabularia* and *Stentor* it has also been shown that premature mitosis can be induced in a post-division nucleus when this is brought into contact with a pre-division cytoplasm.

## Chimaeric cells

Animal viruses, killed by UV irradiation can be used to facilitate cell fusion and hence the production of bi- or multi-nucleate hybrid cells which not only survive for long periods, but, in some cases, multiply. One of the first successful heterokaryons of this kind involved fusion between HeLa cells of human origin and mouse Ehrlich ascites tumour cells. The nuclei of these two cell types are easily distinguished on

morphological grounds and following fusion both sets of nuclei are able
to synthesize DNA and RNA. Evidently the genes of both mice and men
can be replicated and transcribed in cell chimaeras. Heterokaryons can
also be made which involve cells where the synthesis of RNA or DNA or
both is partially or wholly suppressed. In fact Harris has successfully
combined HeLa cells with rabbit macrophages, rat lymphocytes and hen
erythrocytes. All three of these cell types are dormant with respect to
DNA synthesis while the hen erythrocyte is also quiescent in regard to
RNA synthesis. But in hybrid combination with HeLa cells they all
resume the synthesis of those nucleic acids which, under normal circumstances, they do not produce. Thus whenever a cell which lacks the
capacity to synthesize a particular nucleic acid is fused with one which
has that capacity the active partner is able to initiate the appropriate
synthesis in the inactive one. But the reciprocal action is not found. In no
case does the inactive cell type suppress synthesis in the active partner
(Fig. 8.18).

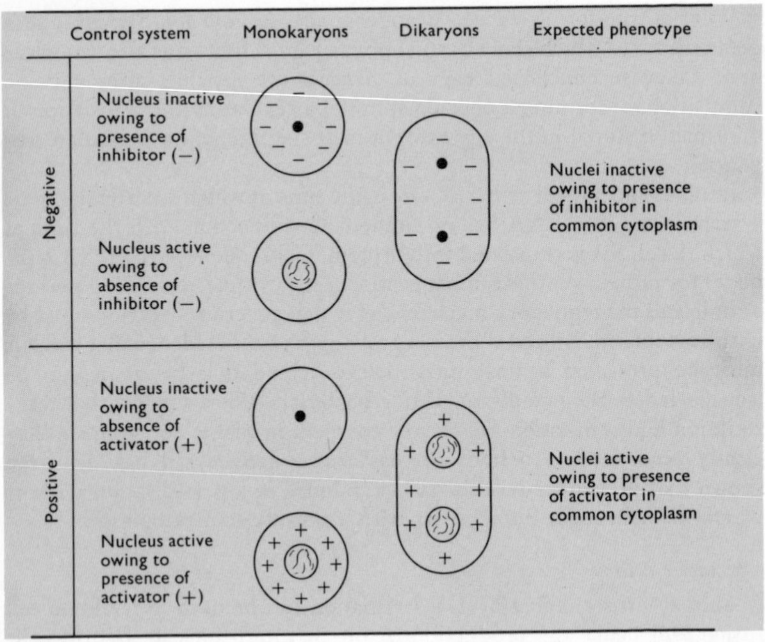

**Fig. 8.18** The phenotypes expected of heterokaryons between active and inactive
cells according to the nature of regulation. Artificially synthesized vertebrate
heterokaryons are active indicating positive control (cf. zygotic induction, immunity to lytic superinfection, etc.).

At the high concentrations in which it is used to promote cell fusion, Sendai virus induces haemolysis in hen erythrocytes and this leads to the disruption and loss of the cell cytoplasm. Consequently the HeLa-hen hybrids are distinctive in that they consist of nuclei from both donor parents but cytoplasm only from the HeLa cells. In other words, they are heterokaryons without being heteroplasmons. The resumption of nucleic acid synthesis in the erythrocyte nuclei must therefore depend upon signals which emanate from HeLa cytoplasm. These signals trigger an enlargement of the hen nuclei accompanied by a dispersion of the condensed chromatin normally present in them.[45]

# 9

# The Organization of Heredity

In earlier chapters we have discussed and exposed units of various kinds. Some of these relate to the genotype in heredity, others to its role in development. Replication, mutation and recombination are the principal features in regard to the hereditary functions of the genotype while transcription, translation and regulation are the essential elements of epigenetic activity.

Viruses and bacteria have but a single genoneme, but cellular organisms, on the other hand, have at least two chromosomes in the haploid complement giving a minimum diploid number of four. This low number is, however, very rare. For example, the average diploid number in flowering plants is about thirty-two while the equivalent number in ferns is over a hundred. In fact, numbers in excess of a thousand have been described in the latter group.

The genoneme and hence the whole genome of viruses and bacteria constitutes a single unit of sequential replication (replicon). But in cellular systems even single chromosome arms are compound in this respect. Indeed, the evidence indicates that, in polytene chromosomes at least, each band, and hence the chromomere from which it arises, represents a unit both in transcription and replication, and over five thousand bands have been identified in the complement of *Drosophila*.

However, studies on amino acid substitution have shown that the smallest unit of independent mutation (muton) is the same in all organisms, namely, a single base.

The unit of recombination (recon), on the other hand, is more difficult to assess. Studies on bacteria show clearly that adjacent base pairs can be separated by crossing-over. This demonstration is valuable from the analytical point of view but, in biological terms, it shows nothing that could not have been inferred. Thus, if a gene consists of an uninterrupted sequence of bases and recombination can occur within the gene, then crossing-over must occur between adjacent bases.[10]

A biologically more meaningful unit would be the average length of the segment between two cross-over points. This can, of course, be estimated from the average number of cross-overs per gene string. This function is given by the total length of the linkage map when this is obtained by adding the observed recombination values for short intervening intercepts. In practice, the total map length is expected to be an

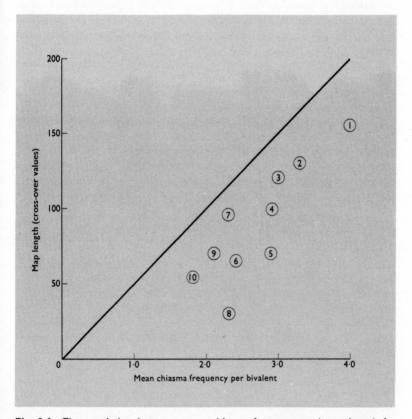

**Fig. 9.1**  The correlation between mean chiasma frequency and map length for the ten chromosomes of *Zea mays* numbered in order of decreasing length. The line shows the maximum map length on the assumption that all chiasmata represent cross-overs and all cross-overs are revealed as chiasmata. Note that the chromosomes differ widely in the degree of marker saturation.

underestimate because even when the density of mutant sites on the map is high, interference is hardly likely to be complete in all the intercepts they define. And unless this condition is satisfied, the observed recombination values do not quite reach the true cross-over values because certain double cross-overs go undetected.

In view of the correlation between chiasma formation and crossing-over, the ratio between chromosome length and cross-over frequency can be determined cytologically in those cellular organisms which are amenable to such analysis. Thus, where $x$ is the chiasma frequency per bivalent, the corresponding cross-over value is $50x$ (Fig. 9.1).

Comparing cross-over rates in higher and lower organisms is somewhat hazardous because the latter can be grown in such large numbers and selective techniques for the detection and isolation of recombinants can be easily applied.

However, it is clear that rates of recombination vary widely between organisms. In $T_4$ phage, for example, one recombination (=map) unit is equivalent to only about 200 base pairs while in *Escherichia* the corresponding value is eight or nine times as large. Consequently, although the physical lengths of the genonemes in these organisms are in the ratio of about $1:20$, their map lengths give a ratio of only $1:2\cdot3$, respectively.

At the other extreme, no crossing-over occurs at all in males of northern species of *Drosophila* (hence the specific reference to females on p. 195) and so the recon is represented here by a whole chromosome.

**Table 9.1** The relationship between genetic and molecular lengths. (Data of Pontecorvo, G. (1959) *Trends in Genetic Analysis,* Oxford University Press, London.)

| Organism | Total length of linkage groups (Map units) | Total amount of DNA per genome (Base pairs) | Base pairs per map unit |
|---|---|---|---|
| Phage $T_4$ | 800 | $1\cdot5 \times 10^5$ | $2\cdot0 \times 10^2$ |
| E. coli | 1,800 | $3 \times 10^6$ | $1\cdot75 \times 10^3$ |
| Aspergillus | 1,000 | $4 \times 10^7$ | $4\cdot0 \times 10^4$ |
| Drosophila | 280 | $8 \times 10^7$ | $3\cdot0 \times 10^5$ |
| Mouse | 1,954 | $5 \times 10^9$ | $3\cdot0 \times 10^6$ |

The fourth unit referred to in regard to the transmission of the genotype was the polaron. But its nature and incidence cannot yet be assessed. Attempts have been made to equate the polaron with units of activity but these have not been acceptable.

Looked at from the point of view of the investigator, this means that what Pontecorvo called the resolving power of the analysis varies greatly between different genetic systems (Table 9.1).

In connection with the epigenetic aspects of gene activity, units of three main kinds were defined, codons, cistrons and operons.

Codons are the three letter words of the non-overlapping, comma-free code which are defined sequentially from the point of translation initiation. Each codon specifies one amino acid or else signals chain termination. The dimensions of the codon and their meanings appear to be universal also and the minor variations which are found can be attributed largely to minor ambiguities and multiple codon recognition.

The cistron can be equated with the structural gene which determines the primary structure of one kind of polypeptide chain. Thus, it can be defined and recognized chemically, on the basis of one-cistron/one-polypeptide chain, or genetically on the grounds that mutations in the same cistron will not give a true wild-type when they are brought together in the trans state. Thus, the cistron, like the muton, appears to be an equivalent unit in all organisms though its detection and definition present problems in certain situations.

Finally, the operon, which typically includes functionally related cistrons, is a unit of co-ordinate transcription, its immediate product being a polycistronic mRNA molecule. Operons are common in bacteria but less so in viruses though genes of related function tend to be grouped together in phage. Mere clustering, however, does not satisfy the diagnostic co-ordinate properties of the operon.[31] In fact, both naturally and by experimental manipulation, genes of unrelated function can be subjected to co-ordinate regulation. Clustering, or partial clustering, has also been found for a number of systems in *Neurospora* and the evidence suggests an operon organization for loci concerned with aromatic amino acid synthesis.

Our analysis goes little beyond this level of genetic organization. It should be appreciated, however, that from a truly biological point of view, subcellular analysis is undertaken with a view to furthering an understanding of cellular and supracellular organization. Thus, what really matters at the biological level is not how an organism effects a process but how that process affects the organism. And the final phenotype is the product of interaction between the environment and the vast variety of products which stem from genetic elements within and without the nucleus.

Further, natural selection adapts the immediate and intermediate phenotype via differential survival and reproduction, which is the outcome of competition between the final, peripheral phenotype and its environment of which other phenotypes form a part.

Indeed, genetic organization does not end with individuals. In sexual systems they collectively constitute different breeding groups within which mating can be regulated in various ways and between which effective mating is impeded or even prohibited. Furthermore, in organisms as different as bees and man the breeding unit and the social organization are themselves related. In the former, the family forms a breeding unit while in man, mating between such close relatives is prohibited by regulations and rules.

Thus, this book is but a bare beginning. But by dealing with the common denominators of genetic organization and its analysis we hope we have prepared the reader for the exploration of wider spheres of genetic influence.

But, above all, we hope to have conveyed the realization that all living organisms are systems of heredity and variation with a common molecular basis and related to each other by processes of heredity and selection.

For too long has the dazzle of Darwinism blinded both biology and society to Mendel's modest discovery.[32] But Genetics has progressed a long way since the days of Johannsen who, having coined the term gene, refused to believe in it and claimed that the word was 'fully free from every hypothesis'.

# Postlude

## THE BIOLOGICAL CONTEXT

Development and reproduction are the principal properties which distinguish the organic and inorganic worlds. These processes and the underlying organization which allows them, clearly arose by processes of selection. Equally clearly, evolutionary response to selection is possible only if the differential development and reproduction on which it rests are themselves inheritable. Thus heredity is both the result of evolution and its indispensable basis. The gene, the chromosome, the cell, the individual, the breeding group and the social organization, are not only evolved units but evolutionary ones. And the dual role of the genotype as controller of development and determiner of heredity makes it and its study central to Biology.

Thus, Genetics aims to elucidate the structure, organization, expression, and transmission of heritable information. It is concerned therefore with a unique means of communication which operates both within the individual during development and between generations during reproduction.

It has also an integrating role to play in the formulation of a natural philosophy and a more obvious one in relation to plant and animal improvement. Indeed, it is not restricted by either functional or taxonomic boundaries, though the latter, insofar as they are natural, are created by the very processes it seeks to study. And if its unifying principles are appreciated and applied, it is relevant to problems of racialism and equality, education and morality, crime and punishment.

Indeed, as long ago as 1900, William Bateson concluded that 'an exact determination of the laws of heredity will probably work more change in man's outlook on the world, and his power over nature, than any other advance in natural knowledge that can be clearly foreseen'. Perhaps this view will prove to be as prophetic as that of Muller's with which we began. Let us hope so.

# Glossary

**Allopatric:** Of two or more populations when their spatial distribution is such as to impede or prohibit interaction between them. The term is usually used in relation to the reproductive isolation of adjacent populations (v. SYMPATRIC).

**Amber mutation:** A suppressible genetic change which creates the chain terminating codon UAG in mRNA.

**Ambiguity:** The phenomenon that a given codon can be translated in more than one way. Used especially in relation to suppressible chain terminating mutations.

**Anticodon:** The particular, specific nucleotide triplet of tRNA which is complementary to the triplet codon of mRNA with which it associates.

**Auxotroph:** A nutritionally-dependent individual or strain whose growth depends on the addition of a supplement to the minimal synthetic medium which will support the growth of nutritionally-independent forms (prototrophs).

**Chiasma interference:** The condition in which one cross-over decreases (positive) or increases (negative) the probability of occurrence of a second cross-over in its vicinity.

**Chromatid interference:** The non-random participation of non-sister chromatids in successive cross-overs.

**Chromomere:** A bead-like concentration of chromosome material owing to localized coiling. The lateral apposition of homologous chromomeres underlies the banded appearance of polytene chromosomes.

**Cistron:** A segment of genetic material which determines the primary structure of a single polypeptide chain. The limits of this functional unit can be determined genetically by means of a cis-trans test.

**Codon:** A nucleotide triplet in the genetic material or the complementary sequence (nodoc) in the corresponding mRNA which codes for a single amino acid or signals the termination of polypeptide chain synthesis.

**Coefficient of coincidence:** The quotient of the number of observed double cross-overs and the number expected on the assumption that single cross-overs occur independently. Coincidence varies inversely with interference.

**Coefficient of integration:** The relationship between the frequency of transfer of a donor marker to recipients and the frequency of its transmission from recipients to recombinants following $F^- \times$ Hfr conjugal mating.

**Dikaryon:** A binucleate state. It represents a third phase in the sexual cycles of ascomycetes and basidiomycetes where it is produced by cell fusion, perpetuated by synchronous division and terminated by nuclear fusion in connection with meiosis. The dikaryon may be homo- or heterokaryotic (v. MONOKARYON).

**Diplophase:** The phase of the life cycle between fertilization and meiosis (v. HAPLOPHASE).

**Endogenote:** That part of the genome of a recipient bacterium which is homologous to the (exogenote) fragment received from a donor in the formation of a merozygote (v. EXOGENOTE).

**Epigenetic:** Relating to the expression, transcription, translation and interaction of the genetic material (cf. GENETIC).

**Episome:** An accessory genetic element which can exist in alternative states, either integrated in a chromosome or as an independent supernumerary entity.

**Euchromatin:** The material of a chromosome, chromosome set or chromosome segment which shows the standard cycle of chromosome coiling (v. HETEROCHROMATIN).

**Exogenote:** v. ENDOGENOTE.

**$F^-$ cells:** Bacterial cells which lack the episomal sex factor (F) and which function as recipients in conjugation (cf. $F^+$ and Hfr CELLS).

**$F^+$ cells:** Bacterial cells which carry the episomal sex factor (F) in its autonomous state and which donate this factor with high frequency during conjugation with $F^-$ recipients thus effecting their conversion to $F^+$ donors (cf. $F^-$ and Hfr CELLS).

**Frame shift mutation:** A deletion or addition of a small number of bases other than three, or multiples thereof, that change the reading frame and, therefore, the spelling of all the codons which lie on the opposite side of the frame shift mutation (reading frame mutation) to the site of translation initiation.

**Genetic:** Relating to the structure, mutation, replication and transmission of the genetic material (cf. EPIGENETIC).

**Genetic compensation:** The condition wherein two recessive mutants, defective in different, functionally non-allelic cistrons (and hence defective for different polypeptides) produce in combination (in mixed infections, heterogenotes, heterokaryons or heterozygotes) a trans dihybrid which is phenotypically wild-type and, therefore, identical to the corresponding cis double heterozygote (cf. GENETIC COMPLEMENTATION).

**Genetic complementation:** The condition wherein two recessive mutants, defective in different ways for the same cistron (and, hence, the same polypeptide) produce in combination (in mixed infections, heterogenotes, heterokaryons or heterozygotes) a trans dihybrid which only approaches the wild phenotype (pseudowild) and which is, therefore, different from the corresponding cis double heterozygote (cf. GENETIC COMPENSATION).

**Genome:** The basic (monoploid or haploid) complement of genetic material in an organism.

**Genoneme:** The gene string of bacteria and viruses which consists of a naked nucleic acid thread.

**Haplophase:** The phase of the life cycle between meiosis and fertilization (v. DIPLOPHASE).

**HeLa cells:** Human cells derived from a carcinoma of the cervix and maintained in tissue culture since 1953. The cell strain is named after Helen Lane, the patient from whom it was obtained.

**Heterochromatin:** The material of chromosomes, chromosome sets or chromosome segments which show heteropycnosity (v. EUCHROMATIN).

**Heteroduplex:** A duplex nucleic acid molecule consisting of two polynucleotide columns which are not completely complementary with regard to base sequence.

**Heterogenote:** Partially diploid (merozygous) bacteria which are heterozygous with regard to loci represented on both the exogenote and the endogenote (v. HOMOGENOTE).

**Heterokaryon:** A bi- or multi-nucleate cell, or a tissue composed of such cells, containing nuclei of more than one genetic type (v. HOMOKARYON). It may constitute the principal vegetative phase in certain fungi.

**Heteroplasmon:** A cytoplasm containing two or more kinds of homologous extrachromosomal genetic elements (v. HOMOPLASMON).

**Heteropycnotic:** Of chromosomes, chromosome sets or chromosome segments which do not follow the standard cycle of coiling and uncoiling and which at any given time may be over (positively heteropycnotic) or under (negatively heteropycnotic) condensed (v. ISOPYCNOTIC).

**Hfr cells:** Bacterial cells carrying the episomal sex factor (F) in its integrated state and which donate (other?) bacterial genes with high frequency to $F^-$ recipients during conjugation (cf. $F^+$ and $F^-$ CELLS).

**Homogenote:** v. HETEROGENOTE.

**Homokaryon:** A bi- or multi-nucleate cell, or a tissue composed of such cells, in which all the nuclei are of one genotype (v. HETEROKARYON).

**Homoplasmon:** v. HETEROPLASMON.

**Isopycnotic:** v. HETEROPYCNOTIC.

**Lysogeny:** The inherent capacity of bacterial cells to produce and release one or more kinds of bacteriophage.

**Lytic cycle:** The multiplication cycle of viruses in their vegetative state leading to the lysis of the host cell.

**Merozygote:** A partially diploid bacterium (homogenote or heterogenote) containing a complete recipient genome and a partial genome (merogenote) derived from a donor.

**Missense mutation:** A base substitution that changes a codon which specifies one amino acid into a codon which specifies a different amino acid.

**Monokaryon:** A uninucleate cell or a tissue composed of such cells (v. DIKARYON).

**Muton:** The smallest mutable unit in the genetic material ($\equiv$ RECON).

**Nonsense mutation:** A base substitution that changes a codon which specifies an amino acid into a codon which signals chain termination.

**Operator:** A region at or near the point of transcription initiation in an operon which is capable of interacting with a specific repressor and, thereby, controlling the reading of the structural genes in the operon.

**Operon:** A segment of the genetic material which functions as a unit of coordinate transcription and specifies a single polycistronic mRNA molecule.

**Paramutation:** A directional, metastable and heritable change in function of one (paramutable) gene which is determined by an allelic (paramutagenic) gene following their combination in a heterozygote.

**Parasexual systems:** Genetic systems which yield recombinants by means other than the regular alternation of fertilization and meiosis.

**Phage conversion:** The acquisition of new characteristics by a bacterium following infection by a phage, the loss of which leads to the disappearance of the acquired characteristics.

**Phenotypic mixing:** The incorporation into the same virus particle of discrepant genome and protein coat specificities owing to the incorporation into the coat of proteins specified by different viral genomes following mixed infection.

**Plaque:** A clear area in a lawn of bacterial cells produced by several cycles of phage infection and lytic multiplication.

**Plasmagene:** An extrachromosomal gene, especially one which cannot be referred to a visible particle.

**Plastogene:** A gene located in a plastid.

**Pleiotropic effects:** Manifold and apparently unrelated changes in the intermediate or peripheral phenotype brought about by a single mutation.

**Polarity:**

  Epigenetic: The phenomenon whereby a (polar) mutation affects not only the polypeptide which corresponds to the cistron in which it occurs but also the rate of synthesis of polypeptides coded by other cistrons which lie on the operator-distal side of the mutated site.

  Genetic: The condition wherein the frequency of gene conversion shown by mutant sites in pair-wise crosses is consistently related to their relative positions ('left' or 'right') in the linear array into which they can be ordered on the linkage map.

**Polaron:** A segment of the genetic material within which mutant sites show a gradient in regard to the frequency with which they undergo gene conversion.

**Polyploidy:** The state in which diplophasic cells, individuals or species have three or more sets of chromosomes.

**Polysomy:** The state in which one or more chromosomes, but not whole sets, are represented three or more times in the diplophase.

**Polyteny:** An increase in the lateral multiplicity of chromosomes brought about by DNA replication without chromatid individualization.

**Position effect:** An alteration of the phenotype owing to a change in relative position of one or more genes or mutational states.

**Prophage:** The integrated state of a temperate phage in which it replicates in concert with the bacterial genome and confers immunity to lytic superinfection on the (lysogenic) bacterium.

**Recon:** The smallest indivisible unit which can be exchanged by recombination ($\equiv$ MUTON).

**Regulator gene:** One that produces an allosteric protein which functions alone (inducible systems) or in combination with a corepressor (repressible systems) in suppressing the transcription of the structural genes in an operon by binding with the operator.

**Replicon:** A unit of sequential replication.

**Sympatric:** Of two or more populations when their spatial distribution is such that they fall, at least partly, within each others' sphere of influence. The term is usually used in relation to reproductive contact (v. ALLOPATRIC).

**Tautomeric shifts:** A reversible change in the position of a proton in a molecule which changes its chemical properties.

**Temperate phage:** An episomal bacterial virus which can exist in alternative states, either integrated non-pathologically into the bacterial genoneme (prophage) and replicating co-ordinately with it, or autonomously in the vegetative state in which it enters the lytic cycle.

**Transduction:** The bacteriophage-mediated transfer of genetic material from a lysed donor to an infected recipient bacterium. The transferred material (exogenote) may become incorporated into the recipient genome (endogenote) and replicate co-ordinately with it (complete transduction) or it may remain as a supernumerary fragment which can function but not replicate (abortive transduction).

**Transformation:** The transfer of genetic information between related bacteria following the release and uptake of naked DNA. Transforming DNA is integrated by replacing the corresponding homologous region of the recipient genome. Integration is a necessary prerequisite for both replication and function (cf. TRANSDUCTION).

**Transition:** The substitution of one purine for another or of one pyrimidine for another at any base location in DNA or RNA (cf. TRANSVERSION).

**Transversion:** The substitution of a purine for a pyrimidine or of a pyrimidine for a purine at any base location in DNA or RNA (cf. TRANSITION).

**Virulent phage:** A bacterial virus for which the vegetative state and the lytic cycle is obligatory (cf. TEMPERATE PHAGE).

**Zygotic induction:** The lysis of a nonlysogenic (ly⁻) recipient (F⁻) bacterium following the transfer of a prophage from a lysogenic (ly⁺) Hfr donor during conjugal mating. Lysis is inevitable under these conditions and depends on the reversion of the prophage to the vegetative state.

# Further Reading and References

## FURTHER READING

BERNAL, J. D. (1967). *The Origin of Life*. Weidenfeld and Nicolson, London.
BONNER, J. (1965). *The Molecular Biology of Development*. Clarendon Press, Oxford.
BRYSON, V. and VOGEL, H. J. (eds) (1965). *Evolving Genes and Proteins*. Academic Press, New York and London.
CARLSON, E. A. (1966). *The Gene: A Critical History*. W. B. Saunders Co., Philadelphia and London.
DAVIDSON, J. N. (1965). *The Biochemistry of the Nucleic Acids*. Methuen and Co., London and New York.
FINCHAM, J. R. S. (1966). *Genetic Complementation*. W. A. Benjamin, Inc., New York and Amsterdam.
FINCHAM, J. R. S. and DAY, P. R. (1965). *Fungal Genetics*, 2nd edn. Blackwell Scientific Publications, Oxford and Edinburgh.
HARRIS, H. (1968). *Nucleus and Cytoplasm*. Clarendon Press, Oxford.
HAYES, W. (1968). *Genetics of Bacteria and their Viruses*, 2nd edn. Blackwell Scientific Publications, Oxford and Edinburgh.
INGRAM, V. M. (1963). *Haemoglobin in Genetics and Evolution*. Columbia University Press, New York and London.
JINKS, J. J. (1964). *Extrachromosomal Inheritance*. Prentice-Hall, Inc., New Jersey.
LEWIS, K. R. and JOHN, B. (1964). *The Matter of Mendelian Heredity*. J. and A. Churchill, London.
MATHER, K. (1938). *The Measurement of Linkage in Heredity*. Methuen and Co., London and New York.
RIEGER, R., MICHAELIS, A. and GREEN, M. M. (1968). *A Glossary of Genetics and Cytogenetics*. Springer-Verlag, Berlin and New York.
SAGER, R. and RYAN, F. J. (1963). *Cell Heredity*. John Wiley and Sons, New York and London.
SCHROEDER, W. A. (1968). *The Primary Structure of Proteins*. Harper and Row, New York and London.

SRB, A. M., OWEN, R. D. and EDGAR, R. S. (1965). *General Genetics*, 2nd edn. W. H. Freeman and Co., San Francisco and London.

SUTTON, H. E. (1962). *Genes, Enzymes and Inherited Disease*. Holt, Rinehart and Winston, New York and London.

WAGNER, R. P. and MITCHELL, H. K. (1964). *Genetics and Metabolism*, 2nd edn. John Wiley and Sons, Inc., New York, London and Sydney.

WATSON, J. D. (1965). *Molecular Biology of the Gene*. W. A. Benjamin, Inc., New York and Amsterdam.

WHITEHOUSE, H. L. K. (1969). *The Mechanism of Heredity*, 2nd edn. Edward Arnold, London.

# REFERENCES

1. ADAMS, J. M. and CAPECCHI, M. R. (1966). N-formyl methionyl-sRNA as the initiator of protein synthesis. *Proc. natn. Acad. Sci. U.S.A.*, **55**, 147–155.
2. ADELBERG, E. A. and PITTARD, J. (1965). Chromosome transfer in bacterial conjugation. *Bact. Rev.*, **29**, 161–172.
3. AMES, B. N. and MARTIN, R. A. (1964). Biochemical aspects of genetics: The operon. *A. Rev. Biochem.*, **33**, 235–258.
4. AMES, B. N. and WHITFIELD, A. J., JNR. (1966). Frameshift mutagenesis in *Salmonella. Cold Spring Harb. Symp. quant. Biol.*, **31**, 221–225.
5. ATTARDI, G. (1967). The mechanism of protein synthesis. *A. Rev. Microbiol.*, **21**, 383–416.
6. AVERY, O. T., MACLEOD, C. M. and MCCARTHY, M. (1944). Studies on the chemical nature of the substance inducing transformation of pneumococcal types. *J. exp. Med.*, **79**, 137–158.
7. BARKSDALE, L. (1959). Lysogenic conversions in bacteria. *Bact. Rev.*, **23**, 202–212.
8. BEADLE, C. W. and TATUM, E. L. (1941). Genetic control of biochemical reaction in *Neurospora. Proc. natn. Acad. Sci. U.S.A.*, **27**, 499–506.
9. BENNELT, J. C. and DREGER, W. J. (1964). Genetic coding for protein structure. *A. Rev. Biochem.*, **33**, 205–234.
10. BENZER, S. (1961). On the topography of genetic fine structure. *Proc. natn. Acad. Sci. U.S.A.*, **47**, 403–415.
11. BONHOEFFER, F., HOSSELBARTH, R. and LEHMANN, K. (1967). Dependence of the conjugal DNA transfer on DNA synthesis. *J. molec. Biol.*, **29**, 539–541.
12. BRADLEY, D. E. (1965). The structure of the head, collar and base-plate of the 'T-even' type bacteriophages. *J. gen. Microbiol.*, **38**, 395–408.
13. BRENNER, S. (1957). On the impossibility of all overlapping triplet codes in information transfer from nucleic acids to proteins. *Proc. natn. Acad. Sci. U.S.A.*, **43**, 687–694.
14. BRENNER, S. (1965). Theories of gene regulation. *Br. med. Bull.*, **21** 244–248.
15. BRETSCHER, M. S. (1968). How repressor molecules function. *Nature, Lond.*, **217**, 509–511.
16. CAIRNS, J. (1964). The chromosomes of *Escherichia coli. Cold Spring Harb. Symp. quant. Biol.*, **28**, 43–46.
17. CALLAN, H. G. (1963). The nature of lampbrush chromosomes. *Int. Rev. Cytol.*, **15**, 1–34.

18. CHAMPE, S. P. (1963). Bacteriophage reproduction. *A. Rev. Microbiol.*, **17**, 87–114.
19. CLINE, A. and BOCK, R. (1967). Translational control of gene expression. *Cold Spring Harb. Symp. quant. Biol.*, **31**, 321–333.
20. CODDINGTON, A. and FINCHAM, J. R. S. (1965). Proof of hybrid enzyme formation in a case of inter-allelic complementation in *Neurospora crassa. J. molec. Biol.*, **12**, 152–161.
21. CRICK, F. H. C. (1966). Codon–anticodon pairing: the wobble hypothesis. *J. molec. Biol.*, **19**, 548–555.
22. CRICK, F. H. C. and BRENNER, S. (1967). The absolute sign of certain phase shift mutants in bacteriophage T₄. *J. molec. Biol.*, **26**, 361–363.
23. CRICK, F. H. C. and ORGEL, L. E. (1964). The theory of interallelic complementation. *J. molec. Biol.*, **8**, 161–165.
24. DEAN, A. C. R. and HINSHELWOOD, C. (1964). What is heredity? *Nature, Lond.*, **202**, 1046–1052.
25. DEMEREC, M. and DEMEREC, Z. (1956). Analysis of linkage relationships in *Salmonella* by transduction techniques. *Brookhaven Symp. Biol.*, **8**, 75–87.
26. DRISKELL-ZAMENHOF, P. (1964). Bacterial episomes, 155–222. In *The Bacteria*, Vol. **5**, GUNSALUS, I. C. and STAINER, R. Y., Academic Press, Inc., New York and London.
27. EMERSON, S. (1948). A physiological basis for some suppressor mutations and possibly for one gene heterosis. *Proc. natn. Acad. Sci. U.S.A.*, **34**, 72–74.
28. EMERSON, S. (1966). Quantitative implications of the DNA-repair model of gene conversion. *Genetics, Princeton*, **53**, 475–485.
29. ENGLESBERG, E., IRR, J., POWER, J. and LEE, N. (1965). Positive control of enzyme synthesis by gene C in the L-arabinose system. *J. Bact.*, **90**, 946–957.
30. FALKOW, S., ROWND, R. and BARON, L. S. (1962). Genetic homology between *Escherichia coli* K-12 and *Salmonella. J. Bact.*, **84**, 1303–1312.
31. FARGIE, B. and HOLLOWAY, B. W. (1965). Absence of clustering of functionally related genes in *Pseudomonas aeruginosa. Genet. Res.*, **6**, 284–299.
32. FISHER, R. A. (1936). Has Mendel's work been rediscovered? *Ann. Sci.*, **I**, 115–137.
33. FRAENKEL-CONRAT, H. L. and WILLIAMS, R. C. (1955). Reconstitution of tobacco mosaic virus from its inactive protein and nucleic acid components. *Proc. natn. Acad. Sci. U.S.A.*, **41**, 690–698.
34. FREESE, E. (1959). On the molecular explanation of spontaneous and induced mutations. *Brookhaven Symp. Biol.*, **12**, 63–75.
35. GAREN, A. and OTSUJI, N. (1964). Isolation of a protein specified by a regulator gene. *J. molec. Biol.*, **8**, 841–852.
36. GARROD, A. E. (1909, 1923). *Inborn errors of metabolism.* Henry Frowde, London.
37. GIERER, A. (1960). Ribonucleic acid as genetic material of viruses. *Symp. Soc. gen. Microbiol.*, **10**, 248–271.
38. GILLIE, O. J. (1966). The interpretation of complementation data. *Genet. Res.*, **8**, 9–31.
39. GOODGAL, S. H. (1961). Studies on transformation of *Hemophilus influenzae*. IV. Linked and unlinked transformations. *J. gen. Physiol.*, **45**, 205–228.

40. GREEN, M. M. and GREEN, K. C. (1956). A cytological analysis of the lozenge pseudoalleles in *Drosophila*. *Z. Vererb. Lehre*, **87**, 708–721.
41. GRIFFITH, F. (1928). Significance of pneumococcal types. *J. Hyg., Camb.*, **27**, 113–159.
42. GROSS, J. G. (1964). Conjugation in bacteria. In *The Bacteria*, Vol. **5**, 1–48. GUNSALUS, I. C. and STAINER, R. Y., Academic Press, Inc., New York and London.
43. GROSS, J. D. and CARO, L. (1966). DNA transfer in bacterial conjugation. *J. molec. Biol.*, **16**, 269–284.
44. HAMMERLING, J. (1963). Nucleo-cytoplasmic interactions in *Acetabularia* and other cells. *A. Rev. Pl. Physiol.*, **14**, 65–92.
45. HARRIS, H. (1967). The reactivation of the red cell nucleus. *J. Cell Sci.*, **2**, 23–32.
46. HARTMAN, P. E. and GOODGAL, S. H. (1959). Bacterial Genetics (with particular reference to genetic transfer). *A. Rev. Microbiol.*, **13**, 465–504.
47. HAWTHORNE, D. C. and MORTIMER, R. K. (1963). Super-suppressors in yeast. *Genetics, Princeton*, **48**, 617–620.
48. HAYASHI, M., HAYASHI, M. N. and SPIEGELMAN, S. (1963). Restriction of genetic *in vivo* transcription to one of the complementary strands of DNA. *Proc. natn. Acad. Sci. U.S.A.*, **50**, 664–672.
49. HAYES, W. (1966). Sex factors and viruses. *Proc. R. Soc.*, B, **164**, 230–245.
50. HAYES, D. (1967). Mechanisms of nucleic acid synthesis. *A. Rev. Microbiol.*, **21**, 369–382.
51. HERSHEY, A. D. and CHASE, M. C. (1952). Independent functions of viral protein and nucleic acid in growth of bacteriophage. *J. gen. Physiol.*, **36**, 39–56.
52. HOLLIDAY, R. (1964). A mechanism for gene conversion in fungi. *Genet. Res.*, **5**, 282–304.
53. HOROWITZ, N. H. and METZENBERG, R. L. (1965). Biochemical aspects of genetics. *A. Rev. Biochem.*, **34**, 527–564.
54. HOTCHKISS, R. D. (1966). Gene, transforming principle and DNA, 180–200. In *Phage and the origins of molecular biology*, CAIRNS, J. J., STENT, G. S. and WATSON, J. D., Cold Spring Harbor Laboratory of Quantitative Biology, Long Island, New York.
55. IMAMOTO, F., MORIKAWA, N. and SATO, K. (1965). On the transcription of the tryptophan operon in *Escherichia coli*. III. Multicistronic messenger RNA and polarity for transcription. *J. molec. Biol.*, **13**, 169–182.
56. ITANO, H. A. and ROBERTSON, E. A. (1960). Genetic control of the α- and β-chains of hemoglobin. *Proc. natn. Acad. Sci. U.S.A.*, **46**, 1492–1501.
57. JACOB, F. and MONOD, J. (1963). Genetic repression, allosteric inhibition and cellular differentiation, 30–64. In *Cytodifferentiation and Macromolecular Synthesis*, *21st Growth Symposium*, Academic Press, Inc., New York.
58. KELLEY, W. S. and SCHAECHTER, M. (1968). The life cycle of bacterial ribosomes. *Adv. Microbiol. Physiol.*, **2**, 89–142.
59. KEYL, H-G. (1965). A demonstrable local and geometric increase in the chromosomal DNA of *Chironomus. Experientia*, **21**, 191–193.
60. LACKS, S. (1962). Molecular fate of DNA in genetic transformation of *Pneumococcus. J. molec. Biol.*, **5**, 119–131.
61. LACKS, S. (1966). Integration efficiency and genetic recombination in pneumococcal transformation. *Genetics, Princeton*, **53**, 207–235.

62. LARK, K. G. (1966). Regulation of chromosome replication and segregation in bacteria. *Bact. Rev.*, **30**, 3–32.

63. LEWIS, E. B. (1949). The phenomenon of position effect. *Adv. Genet.*, **3**, 73–115.

64. LODISH, H. F. and ZINDER, N. D. (1966). Replication of the RNA of bacteriophage f2. *Science, N.Y.*, **152**, 372–378.

65. LYON, M. F. (1963). Attempts to test the inactive-X theory of dosage compensation in mammals. *Genet. Res.*, **4**, 93–103.

66. MESELSON, M. S. and STAHL, F. W. (1958). The replication of DNA in *Escherichia coli*. *Proc. natn. Acad. Sci. U.S.A.*, **44**, 671–682.

67. MILLER, I. L. and LANDMAN, O. E. (1966). On the mode of entry of transforming DNA into *Bacillus subtilis*, 187–194. In *The physiology of gene and mutation expression*, KOHOUTOVA, M. and HUBACEK, J., Academia, Prague.

68. NIRENBERG, M. W. and MATTHAEI, J. H. (1961). The dependence of cell-free protein synthesis in *Escherichia coli* upon naturally occurring or synthetic polyribonucleotides. *Proc. natn. Acad. Sci. U.S.A.*, **47**, 1588–1602.

69. PERRY, R. P. (1966). On ribosome biogenesis. *Natn. Cancer Inst. Monogr.*, **23**, 527–545.

70. PTASHNE, M. (1967). Specific binding of the λ phage repressor to DNA. *Nature, Lond.*, **214**, 232–234.

71. RAVIN, A. W. (1961). The genetics of transformation. *Adv. Genet.*, **10**, 61–163.

72. RICH, A., WARNER, J. R. and GOODMAN, H. M. (1963). The structure and function of polyribosomes. *Cold Spring Harb. Symp. quant. Biol.*, **28**, 269–285.

73. RICHMOND, M. H. (1968). The plasmids of *Staphylococcus aureus* and their relation to other extrachromosomal elements in bacteria. *Adv. Microbiol. Physiol.*, **2**, 43–88.

74. RITOSSA, F. M. and SPIEGELMAN, S. (1965). Localization of DNA complementary to ribosomal RNA in the nucleolus organizer region of *Drosophila melanogaster*. *Proc. natn. Acad. Sci. U.S.A.*, **53**, 737–745.

75. SADGOPAL, A. (1968). The genetic code after the experiment. *Adv. Genet.*, **14**, 326–404.

76. SAGER, R. and RAMANIS, Z. (1965). Recombination of nonchromosomal genes in *Chlamydomonas*. *Proc. natn. Acad. Sci. U.S.A.*, **53**, 1053.

77. SALAS, M., SMITH, M., STANLEY, W. M. JNR., WAHBA, A. and OCHOA, S. (1965). Direction of reading of the genetic message. *J. biol. Chem.*, **240**, 3988–3995.

78. SARABHI, A. S., STRETTON, A. O. W., BRENNER, S. and BOLLE, A. (1964). Colinearity of the gene with the polypeptide chain. *Nature, Lond.*, **201**, 13–17.

79. SCAIFE, J. (1967). Episomes. *A. Rev. Microbiol.*, **21**, 601–638.

80. SCHAEFFER, P. (1964). Transformation. In *The Bacteria*, Vol. **5**, 87–153, GUNSALUS, I. C. and STAINER, R. Y., Academic Press, Inc., New York and London.

81. SCHLESINGER, M. J. and LEVINTHAL, C. (1965). Complementation at the molecular level of enzyme interaction. *A. Rev. Microbiol.*, **19**, 267–284.

82. SCOTT, N. S. and SMILLIE, R. M. (1967). Evidence for the direction of chloroplast ribosomal RNA synthesis by chloroplast DNA. *Biochem. biophys. Res. Commun.*, **28**, 598–603.

83. SKALKA, A. (1968). Regional and temporal control of genetic transcription in phage lambda. *Proc. natn. Acad. Sci. U.S.A.*, **55**, 1190–1195.

84. SPIZIZEN, J., REILLY, B. E. and EVANS, A. H. (1966). Microbial transformation and transfection. *A. Rev. Microbiol.*, **20**, 371–400.

85. STAHL, F. W., EDGAR, R. S. and STEINBERG, J. (1964). The linkage map of bacteriophage T₄. *Genetics, Princeton*, **50**, 539–552.

86. STANIER, R. Y. (1951). Enzymic adaptation in bacteria. *A. Rev. Microbiol.*, **5**, 35–56.

87. TOMASZ, A. (1965). Control of the competent state in *Pneumococcus* by a hormone-like cell product: an example of a new type of regulatory mechanism in bacteria. *Nature, Lond.*, **208**, 155–159.

88. UMBARGER, H. E. (1964). Intracellular regulatory mechanisms. *Science, N.Y.*, **145**, 674–679.

89. VISCONTI, N. (1966). Mating theory, 142–149. In *Phage and the origins of molecular biology*, CAIRNS, J., STENT, G. S. and WATSON, J. D., Cold Spring Harbor Laboratory of Quantitative Biology, Cold Spring Harbor, Long Island, New York.

90. VOGEL, H. J. and VOGEL, R. H. (1967). Regulation of protein synthesis. *A. Rev. Biochem.*, **36**, 519–538.

91. WATSON, J. D. and CRICK, F. H. C. (1953). The structure of DNA. *Cold Spring Harb. Symp. quant. Biol.*, **18**, 123–131.

92. WEIGERT, M. G., LANKA, E. and GAREN, A. (1965). Amino acid substitutions resulting from suppression of nonsense mutations. II. Glutamine insertion by the Su-2 gene; tyrosine insertion by the Su-3 gene. *J. molec. Biol.*, **14**, 522–527.

93. WEISSMANN, C. and OCHOA, S. L. (1967). Replication of phage RNA. *Prog. nucl. acid Res. molec. Biol.*, **6**, 353–399.

94. WHITEHOUSE, H. L. K. (1965). A theory of crossing-over and gene conversion involving hybrid DNA. *Proc. 11th Int. Conf. Genet.*, 1963, **2**, 87–88.

95. WILKINS, M. H. F. (1961). The molecular structure of DNA. *J. Chimie phys.*, **58**, 891–898.

96. YANOFSKY, C. (1960). The tryptophane synthetase system. *Bact. Rev.*, **24**, 221–245.

97. YANOFSKY, C., CARLTON, B. C., GUEST, J. R., HELINSKI, D. R. and HENNING, U. (1964). On the colinearity of gene structure and protein structure. *Proc. natn. Acad. Sci. U.S.A.*, **51**, 266–272.

98. YOSHIKAWA, H. and SUEOKA, N. (1963). Sequential replication of *Bacillus subtilis* chromosome. I. Comparison of marker frequencies in exponential stationary growth phases. *Proc. natn. Acad. Sci. U.S.A.*, **49**, 559–566.

99. ZICHICHI, M. L. and KELLENBERGER, G. (1963). Two distinct functions in the lysogenisation process: the repression of phage multiplication and the incorporation of the prophages in the bacterial genome. *Virology*, **19**, 450–460.

100. ZINDER, N. D. (1965). RNA phages. *A. Rev. Microbiol.*, **19**, 455–472.

# Classified Species Index

# General Index

74

71 673 S    N 86

## ABOUT THIS BOOK

Heredity is the essence of life and an understanding of it is crucial in the study of a wide range of biological and para-biological problems.

Little advance was made towards elucidating the phenomena of heredity and variation until Mendel's conclusions were confirmed, extended and publicly debated. However, the advances of the last twenty years have transformed the subject completely.

This book emphasizes recent advances in molecular and microbial genetics at a level appropriate to an introductory course in these important and advancing areas of enquiry. The methods used for mapping mutant sites in various organisms are considered in some detail because they are relevant to the investigation of numerous aspects of genetic organization.

This account of the molecular and analytical approach to the study of heredity will enable the reader to realize the essential unity of life and to apply the principles revealed to the study of higher levels of biological organization.

American Elsevier